T0350904

Electrical Systems for Nuclear Power Plants

Electrical Systems for Nuclear Power Plants

Omar S. Mazzoni

Published by
Standards Information Network

IEEE PRESS

WILEY

This edition first published 2019
© 2019 by The Institute of Electrical and Electronics Engineers, Inc.

The rights of Omar S. Mazzoni to be identified as the author of this work have been asserted in accordance with law.

Registered Office
John Wiley & Sons, Inc., 111 River Street, Hoboken, NJ 07030, USA

Editorial Office
111 River Street, Hoboken, NJ 07030, USA

For details of our global editorial offices, customer services, and more information about Wiley products visit us at www.wiley.com.

Wiley also publishes its books in a variety of electronic formats and by print-on-demand. Some content that appears in standard print versions of this book may not be available in other formats.

Library of Congress Cataloging-in-Publication Data

ISBN: 9781119483601

Cover design: Wiley
Cover image: © MichaelUtech/GettyImages

Set in 10/12pt WarnockPro by Aptara Inc., New Delhi, India

Printed in the United States of America.

V10004414_091018

This book is dedicated to my beloved daughter, Ruth Maria Mazzoni, who is ever present in my mind and whom I am missing greatly since her early departure from this life on January 23, 2012.

Contents

Preface

This book is the result of class notes for a graduate-level course in electrical engineering, taught by the author at The George Washington University, at several times in the period 1990–2017. The book includes problems and questions which were posed to the students to test their understanding of the material given in the class and to stimulate class discussions. The main reason for the book has been the lack of other suitable instructional material on the subject.

The book is intended for a graduate-level university course in power engineering programs. Also, the book should be useful for training of beginning engineers and as a ready reference for the practicing engineer.

The book summarizes the author's experience in the grass roots design of nuclear generating stations, began at Burns and Roe, Inc. from 1969 to 1979, where left as Assistant Chief Electrical Engineer. Also important was the experience gained in the modifications to nuclear generating stations, while working as Manager of Electrical and Instrumentation Systems at NUS Corporation, from 1983 to 1989, and the experience gained in performing inspections and plant evaluations for numerous nuclear utility clients, with Systems Research International, Inc. (SRI), of which he is the principal founder-owner. The author has also performed 51 inspections at nuclear plants, as an electrical and instrumentation consultant for the US Nuclear Regulatory Commission over the period 1989–2016.

Throughout his technical career, the author has been very active in the development of standards at the Institute of Electric and Electronics Engineers, Inc. (IEEE), participating in standards working groups. The author remains very active in the balloting of IEEE standards. IEEE standards related to nuclear plants were obviously a fundamental guide for this book.

Numerous sources were consulted in the development of the book material, but the author is very thankful to the following institutions in particular:

- US Nuclear Regulatory Commission,
- Institute of Electric and Electronics Engineers, Inc.,

- Westinghouse Electric Corporation, and
- The George Washington University.

Of the many references given in the book, only the edition cited applies for dated references. For undated references, the latest edition of the referenced document (including any amendments and corrigenda) applies.

The author wishes to thank The George Washington University for providing him the opportunity to teach the course. Also my thanks go to the many students who took the course through the years, to the Department of Electrical Engineering and Computer Science, and especially to Professor Robert Harrington, who was a friendly guiding source. The author also has heartfelt gratitude for the encouragement and patience of his wife Maria (Betty) during the writing of the book.

It is hoped that the book may be usable to professionals interested in understanding and teaching the principal criteria for nuclear power plant electrical and instrumentation systems.

Professor Omar S. Mazzoni, DSc, PE

1

Elements of a Power System

This chapter discusses the main elements of a power system that are particularly applicable to understanding the basic electrical configuration of a nuclear power plant.

1.1 The Alternating Current One-Line Diagram

A useful diagram to depict the alternating current (ac) system of a nuclear generating station is the one-line diagram (OLD) [2]. The OLD allows for the simplest representation of the main elements of a typical station and the interconnections among them. Figure 1.1 depicts a typical OLD for the high voltage and the medium voltage portions of a nuclear power station. Depending on the plant design and year of start of operations, the OLD may be somewhat different than the one depicted in Figure 1.1. The OLD of Figure 1.1 depicts a preferred approach to nuclear station design, as it offers enhanced independence for the safety-related busses.

While transformers AT1 and AT2 can be connected to the safety buses and also to the nonsafety buses, they are mainly intended to feed the safety-related busses. The only time that AT1 and AT2 are used to supply the plant nonsafety-related loads is for plant start-up or shutdown.

The plant normal operating conditions are with transformer AT feeding the plant nonsafety-related loads. Transformers AT1 and AT2 and normally energized and feed the safety buses, with both of these transformers being assigned to the two safety-related buses. Each one of the safety-related buses can be fed from an emergency diesel generator (EDG). Under normal plant operation, the EDGs are on standby condition.

After the generator starts up and synchronized to the grid, it is initially loaded with grid load to about 30% normal plant load, at which time the secondary side breakers of transformer AT are closed to feed the plant auxiliary loads.

Electrical Systems for Nuclear Power Plants, First Edition. Omar S. Mazzoni.
© 2019 by The Institute of Electrical and Electronic Engineers, Inc. Published 2019 by John Wiley & Sons, Inc.

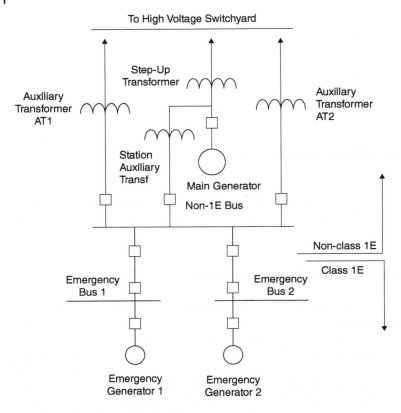

Figure 1.1 One-line diagram of nuclear station.

The preferred mode of starting up the station would be by synchronizing across the generator circuit breaker and feeding the station auxiliaries through transformer AT. An alternate approach, though less preferred, would be through either transformer AT1 or AT2, which would require a bus transfer to eventually feed plant loads through transformer AT.

Automatic bus transfers are necessary to prevent paralleling of two sources, which would impose undue short circuit stresses for the circuit breakers.

Thorough the plant start-up process, the two safety-related buses remain energized and are not subjected to any transfer operations, thereby eliminating the possibility of transients to be induced in the safety system.

The plant design allows for the possibility of testing the EDGs as required by the plant surveillance. The EDGs are tested once a month, and loads are picked up from the system for the period of test duration.

1.2 Basis for One-Line Representation

For a three-phase power system, in order properly represent it in an OLD, the system must be balanced, that is each one of the three phases must carry the same magnitude of current, and the currents must be at 120° from each other. As a result, the balanced three-phase system may be represented by just one of its phases, as they are the same to each other.

Percent and Per Unit Representation

In the definition of electrical power system values, a representation in percent (%) or per unit (pu) is more convenient than actual units such as ohm, volts, etc,

A base value must be introduced to show quantities in percent or per unit. For example, assuming a base voltage of 4.16 kV:

4.0 kV becomes 0.9615 pu, 96.15%
7.2 kV becomes 1.73 pu, 173%

The pu system is preferred because the product of two pu values results in pu, whereas the product of two percent values must be divided by 100 to express the result in percent.

Two base values must be selected to have all other values automatically determined. Generally, kVA and kV are selected. The term MVA refers to MVA for three phase and the term kV to voltage from line to line, then the term current refers to line current. Generally, power is understood to be three-phase power and voltage the line-to-line voltage, unless otherwise specified. A major exception occurs in the method of symmetrical components, where line-to-neutral voltage is used.

Per unit and percent values of transformer impedances are the same on either side of the unit (primary or secondary). Also, the per unit and percent values do not depend on the transformer connections (wye–wye, delta–wye…etc.)

The following relationships apply (see [4]):

Base current, A = base kVA/1.732 base kV
Ω = (base voltage, kV$_{LN}$)^2/base MVA
Base impedance, Ω = (base voltage, kV$_{LN}$)^2/base MVA
Per unit impedance = actual impedance, Ω/base impedance, Ω

When dealing with unbalanced systems, symmetrical components must be used. In this method, the sequence components are always line to neutral or line to ground, as appropriate.

1.3 Main Electrical Components of Power Plants

Rotating Machines

Generators
Two types of generators are found in nuclear power plants: main generators that convert the mechanical power into electrical power and emergency standby generators utilized to provide electrical power to safety-related equipment when the normal power is not available [1,3]. This course places particular emphasis on the review of requirements for standby generators, as they are intrinsically involved in the performance of safety-related functions.

Motors
Most motors encountered in safety-related systems for nuclear power plants are of the AC induction type, constant speed. Some plants have AC variable speed drives, such as the feed water pumps for boiling water reactor type reactors. Motors for reactor cooling pumps are generally fitted with flywheels, to extend their operation when power is lost. Motors fitted with flywheels present special starting and protection issues.

DC motors are utilized for safety-related motor-operated valves.

Transformers

Distribution transformers are rated from 3 to 500 kVA, and below power transformers are those rated above 500 kVA.

In terms of insulation type, transformers can be oil cooled or dry type. The dry-type transformers include ventilated, cast coil, totally enclosed nonventilated, and sealed gas filled. Oil-insulated transformers are not utilized indoors, due to their intrinsic fire hazard.

Cables

Cables can be classified in accordance with their insulation ratings, which are as follows:

- medium voltage, rated 2001–35,000 V,
- low voltage, generally rated at 600 V, and
- instrumentation and control, rated below 600 V.

Switchgear, Distribution Panelboards, and Motor Control Centers

- switchgear, medium voltage, rated above 2000 V,
- switchgear, low voltage, rated 600 V, and
- panelboards and motor control centers, rated 600 V and below.

Buses

- *Phase-segregated buses*: Each phase is enclosed in a metallic enclosure, allowing for better isolation. Generally utilized for higher voltage types, such as the generator connection to the step-up transformer (19–24 kV).
- *Nonsegregated buses*: All phases are in the same metallic enclosure, generally utilized for voltages at 7.2–4.16 kV.

DC Batteries

Generally lead acid are the types of batteries found in nuclear plants. Voltages are between 24 and 250 V.

Battery chargers are provided to keep the batteries fully charged. The chargers provide the DC system load requirements whenever the normal ac input to the charger is available. Whenever ac power is available, the chargers provide the normal DC load and also the "trickle" current necessary to keep the battery fully charged.

Inverters

Inverters provide 120–240 V ac to safety loads requiring power when the main ac is not available, normally provided with autotransfer capabilities between available power sources.

Plant Loads and Their Characteristics

It is important to know how the plant loads will behave under variable voltage conditions. The modeling of plant loads is performed in accordance with the following guidelines:

- motors, constant impedance type for starting conditions, constant kVA type for running conditions and
- transformers, constant impedance type
  - lighting, load varies as the square of the impressed voltage,
  - heaters, load varies as the square of the impressed voltage, and
  - inverters constant kVA.

1.4 Transmission Lines, Switchyards, and Substations

The switchyard and associated substation are required to connect the power plant to the transmission grid, but are under the control of the power plant operators. The switchyard and its associated breakers, lines, and components are not considered safety related, though they may be considered important to safety.

Strategic arrangements are provided to increase reliability of operation. One common arrangement, which provides very high reliability, consists of three breakers in on lineup with the end breakers connected to a separate bus and two lines between the three breakers (this is the so-called "breaker-and-a-half" scheme).

Operation: Transmission lines are operated by the regional associations of grid operators. Of particular importance to the nuclear plant is the regulation of the supplied voltage, as it will affect the voltage at the plant and the operation of safety-related equipment. Special voltage relays are provided to disconnect the plant from the transmission grid, should the voltage excursion become intolerable for the proper operation of the safety equipment in the plant. (See [5] for extreme external events). The frequency of the transmission grid is also important, particularly when an EDG is paralleled to grid for test purposes.

Transmission Line Protective Relaying

Transmission line protective relaying is discussed under Chapter 9, "Interface of the Nuclear Plant and the Grid."

Transmission Line Testing and Inspection

Transmission testing and inspection is discussed under Chapter 9, "Interface of the Nuclear Plant and the Grid."

Questions and Problems

1.1 A prime requirement for the application of the one-line diagram is that the three-phase power system could be represented by just one phase. Could this representation be possible for a distribution system where there are numerous loads between one phase and the neutral? Explain your answer.

1.2 If a power system develops a phase-to-phase fault, is it possible to represent the faulted system with a one-line diagram? Explain your answer

1.3 What analytical technique is available to study unbalanced system conditions in a three-phase power circuit?

1.4 The one-line diagram (OLD) is utilized in the nuclear plant control room by the plant operators to review plant conditions under normal and emergency operation. In addition, a plant simulator is provided. The simulator is mainly used for training purposes. If there is a discrepancy between the simulator and the OLD, which one takes precedence? Explain.

1.5 A main generator circuit breaker is included in modern nuclear plant designs. Other plants have the generator connected directly to the switchyard through the step-up transformer. The main generator circuit breaker allows for isolation of the generator from the switchyard and also allows for availability of the plant station service transformer to feed the plant auxiliaries, without the need for fast bus transfer scheme. Would a generator breaker be considered particularly important to safe shutdown of the plant? Explain.

1.6 Plants that do not have a generator circuit breaker usually incorporate a fast bus transfer scheme to continue powering the station auxiliary buses in case of a plant trip. Why does the bus transfer scheme need to be fast? Provide a brief explanation.

1.7 From description of plant operation, it is not possible to keep the plant buses energized from two sources, as the short circuit capability of the breakers would be exceeded. Is this a normal plant design approach? If so, why?

1.8 If a nuclear plant operating at 100% output gets separated from the grid for a period time, say, for example, 5 min, it will need to trip and shutdown. This is due to the large excess between the reactor energy and the energy represented by the turbine steam bypass plus the plant auxiliary loads (about 6% of full load). The turbine bypass of the reactor energy is typically 25%. This situation is typical of all nuclear plants operating in the United States.

What if anything could be done to allow the plant to ride over the transient and keep the generator and reactor running until reconnection to the grid becomes possible?

1.9 Given:

Line to line voltage $= 7.2$ kV

Line to neutral voltage $= 7.2/\sqrt{3} = 4/16$ kV

Three phase power $= 21,000$ kVA

$kVA_{1\Phi} = 7000$ kVA

Calculate pu values.

1.10 Later designs of nuclear plants have been fitted with a main generator circuit breaker, and some existing plants have been retrofitted to incorporate generator circuit breakers. Comment on the following:
- Have generator circuit breakers enhanced the operation of nuclear plants?, if so how?
- What safety aspects are enhanced by the generator circuit breaker?

References

Publishing organization websites:

US Nuclear Regulatory Commission: https://www.nrc.gov/

Institute of Electrical and Electronics Engineers: www.ieee.org

International Atomic Energy Agency: www.iaea.org

1 IEEE 100 Std., "Dictionary of Electrical and Electronic Terms," Sixth ed., 1996.

2 US NRC NUREG-0544. "Collection of Abbreviations."

3 D. G. Fink and H. Beaty, *Standard Handbook for Electrical Engineers* McGraw-Hill, New York, 2007.

4 IEEE 141, "Recommended Practice for Electric Power Distribution for Industrial Plants."

5 "Assessment of Vulnerabilities of Operating Nuclear Power Plants to Extreme External Events," IAEA TECDOC No. 1834, 2017.

2

Nuclear Power Plants: General Information

2.1 Introduction

This chapter provides a brief review of nuclear power generation, which is necessary to study the electrical systems and their importance to safety. Nuclear plant accidents are discussed with particular emphasis on the accident contribution from electrical and instrumentation safety systems operation.

2.2 Environmental Impact

Fossil fuel generating plants differ from nuclear plants in that the steam supply system consists of boilers, where the heat is supplied by a conventional fossil fuel such as oil, coal, or gas. Burning of fossil fuel creates difficult environmental issues. Nuclear plants, on the other hand, do not create environmental issues, unless safety of the plants is compromised to the point of creating nuclear accidents. Major nuclear accidents have occurred, and some have had a profound impact on the environment, such as Chernobyl and Fukushima.

On the record of the past 50 years, nuclear power has an edge over other forms of providing energy both in terms of limiting day-to-day adverse health and environmental effects, including greenhouse gas emissions and in terms of the frequency and toll of major accidents. Table 2.1 makes this point clear.

The low morbidity is due to several factors, but two stand out:

- Most casualties and other health and environmental effects stem from the extractive and transportation industries. Because the same amount of electric power can be obtained from about 200–300 tons of uranium ore as from 3 to 4 million tons of coal or similarly large quantities of gas or oil, these effects are inherently less severe for nuclear power than for the main hydrocarbon sources of electricity.
- Safety culture needs to be enforced and strengthened at every nuclear generating plant, so that there will be no more accidents at nuclear power generating facilities.

Electrical Systems for Nuclear Power Plants, First Edition. Omar S. Mazzoni.
© 2019 by The Institute of Electrical and Electronic Engineers, Inc. Published 2019 by John Wiley & Sons, Inc.

Table 2.1 Main Sources of Electricity in the World and Their Morbidity and Greenhouse Gas Emissions Per Unit of Electricity Produced

Source (Percentage of world use, 2007)	Deaths per terawatt-hour	Tons of greenhouse gas emissions per gigawatt-hour (life)
Nuclear (14%)	0.04	Small–50
Coal (42%)	161 (US average is 15)	800–1400
Gas (21%)	4	300–500
Hydro (16%)	0.1 (Europe)	Small–100
Wind (< 1%)	0.15	Small–50

Table generated by authors using data from Key World Energy Statistics 2009[4].

2.3 Nuclear Generation Fuel Cycle

Mining and Milling

Uranium is usually mined by either surface (open cut) or underground mining techniques, depending on the depth at which the ore body is found. In Australia, the Ranger mine in the Northern Territory is open cut, whereas Olympic Dam in South Australia is an underground mine (which also produces copper, with some gold and silver). The newest Canadian mines are underground.

From these, the mined uranium ore is sent to a mill where the ore is crushed and ground to a fine slurry, which is leached in sulfuric acid to allow the separation of uranium from the waste rock. It is then recovered from solution and precipitated as uranium oxide (U_3O_8) concentrate (sometimes this is known as "yellowcake," though it is finally khaki in color).

U_3O_8 is the uranium product which is sold. About 200 tonnes is required to keep a large (1000 megawatt electric [MWe]) nuclear power reactor generating electricity for 1 year.

Conversion

Because uranium needs to be in the form of a gas before it can be enriched, the U_3O_8 is converted into the gas uranium hexafluoride (UF_6) at a conversion plant.

Enrichment

The vast majority of all nuclear power reactors in operation and under construction require "enriched" uranium fuel in which the proportion of the U-235 isotope has been raised from the natural level of 0.7% to about 3.5–5%. The enrichment process removes about 85% of the U-238 by separating gaseous uranium hexafluoride into two streams: One stream is enriched to the required level and then passes to the next stage of the fuel cycle. The other stream is depleted in U-235 and is called "tails." It is mostly U-238.

So little U-235 remains in the tails (usually less than 0.25%) that it is of no further use for energy, though such "depleted uranium" is used in metal form in yacht keels, as counterweights, and as radiation shielding, since it is 1.7 times denser than lead.

Fuel Fabrication

Enriched UF_6 is transported to a fuel fabrication plant where it is converted to uranium dioxide (UO_2) powder and pressed into small pellets. These pellets are inserted into thin tubes, usually of a zirconium alloy (zircalloy) or stainless steel, to form fuel rods. The rods are then sealed and assembled in clusters to form fuel assemblies for use in the core of the nuclear reactor.

Some 27 tonnes of fresh fuel is required each year by a 1000-MWe reactor.

Several hundred fuel assemblies make up the core of a reactor. For a reactor with an output of 1000 MWe, the core would contain about 75 tonnes of low-enriched uranium. In the reactor core, the U-235 isotope fissions or splits, producing heat in a continuous process called a chain reaction. The process depends on the presence of a moderator such as water or graphite and is fully controlled.

Some of the U-238 in the reactor core is turned into plutonium, and about half of this is also fissioned, providing about one third of the reactor's energy output.

As in fossil fuel burning electricity generating plants, the heat is used to produce steam to drive a turbine and an electric generator, in this case producing about 7 billion kilowatt hours of electricity in 1 year.

To maintain efficient reactor performance, about one third of the spent fuel is removed every year or 18 months, to be replaced with fresh fuel.

Used Fuel Storage

Used fuel assemblies taken from the reactor core are highly radioactive and give off a lot of heat. They are therefore stored in special ponds which are usually located at the reactor site, to allow both their heat and radioactivity to decrease. The water in the ponds serves the dual purpose of acting as a barrier against radiation and dispersing the heat from the spent fuel.

Used fuel can be stored safely in these ponds for long periods. It can also be dry stored in engineered facilities, cooled by air. However, both kinds of storage are intended only as an interim step before the used fuel is either reprocessed or sent to final disposal. The longer it is stored, the easier it is to handle, due to decay of radioactivity.

There are two alternatives for used fuel:

1. reprocessing to recover the usable portion of it and
2. long-term storage and final disposal without reprocessing.

Reprocessing

Used fuel still contains approximately 96% of its original uranium, of which the fissionable U-235 content has been reduced to less than 1%. About 3% of used fuel comprises waste products, and the remaining 1% is plutonium (Pu) produced while the fuel was in the reactor and not "burned" then.

Reprocessing separates uranium and plutonium from waste products (and from the fuel assembly cladding) by chopping up the fuel rods and dissolving them in acid to separate the various materials. Recovered uranium can be returned to the conversion plant for conversion to uranium hexafluoride and subsequent re-enrichment. The reactor-grade plutonium can be blended with enriched uranium to produce a mixed oxide (MOX) fuel, in a fuel fabrication plant.

MOX fuel fabrication occurs at facilities in Belgium, France, Germany, United Kingdom, Russia, and Japan, with more under construction. There have been 25 years of experience in this, and the first large-scale plant, Melox, commenced operation in France in 1995. Across Europe, about 30 reactors are licensed to load 20–50% of their cores with MOX fuel and Japan plans to have one third of its 54 reactors using MOX by 2010.

The remaining 3% of high level radioactive wastes (some 750 kg per year from a 1000-MWe reactor) can be stored in a liquid form and subsequently solidified.

Reprocessing of used fuel occurs at facilities in Europe and Russia with capacity over 5000 tonnes per year and cumulative civilian experience of 90,000 tonnes over almost 40 years.

Vitrification

After reprocessing, the liquid high level waste can be calcined (heated strongly) to produce a dry powder, which is incorporated into borosilicate (Pyrex) glass to immobilize the waste. The glass is then poured into stainless steel canisters, each holding 400 kg of glass. A year's waste from a 1000-MWe reactor is contained in 5 tonnes of such glass, or about 12 canisters 1.3 m high and 0.4 m in diameter. These can be readily transported and stored with appropriate shielding.

This is as far as the nuclear fuel cycle goes at present. The final disposal of vitrified high-level wastes, or the final disposal of spent fuel, which has not been reprocessed used fuel, has not yet taken place.

Final Disposal

The waste forms envisaged for disposal are vitrified high-level wastes sealed into stainless steel canisters, or used fuel rods encapsulated in corrosion-resistant metals such as copper or stainless steel. All national policies intend either kind of canisters to be buried in stable rock structures deep underground. Many geological formations such as granite, volcanic tuff, salt, or shale are suitable. The first permanent disposal is expected to occur about 2020 (this date may still be questionable).

2.4 Evolution of Nuclear Power Generation

1. Early reactors – year 1950
2. Current reactors – year 1970
3. Evolutionary reactors – year 2000
4. Advance ecolutionary and passive reactors – year 2020
5. Other advanced technologies, including hydrogen generation

2.5 Nuclear Power in the United States

As of 2016, nuclear power in the United States is provided by 100 commercial reactors (66 pressurized water reactors (PWR) and 34 boiling water reactors) licensed to operate at 61 nuclear power plants, producing a total of 797.2 terawatt-hours of electricity, which accounted for 19.50% of the nation's total electric energy generation in 2015. As of 2016, there were four new reactors under construction with a gross electrical capacity of 5000 MW. The United States is the world's largest supplier of commercial nuclear power, and in 2013 generated 33% of the world's nuclear electricity.

In recent developments, there has been some revival of interest in nuclear power in the 2000s, with talk of a "nuclear renaissance," supported particularly by the Nuclear Power 2010 Program. A number of applications was sought, and construction on a handful of new reactors began in the early 2010s, in late 2011 and early 2012, construction of four new nuclear reactor units at two existing plants were approved, the first such in 34 years. However, facing economic challenges, and later in the wake of the 2011 Japanese nuclear accidents, most of these projects have been canceled, and as of 2012, "nuclear industry officials say they expect just five new reactors to enter service by 2020—Southern's two Vogtle reactors, two at Summer in South Carolina and one at Watts Bar in Tennessee"; these are all at existing plants.

2.6 Plans for New Reactors Worldwide

Nuclear power capacity worldwide is increasing steadily, with over 60 reactors under construction in 13 countries.

2.7 Increased Capacity

Increased nuclear capacity in some countries is resulting from the uprating of existing plants. This is a highly cost-effective way of bringing on new capacity.

2.8 Nuclear Plant Construction

In all, about 160 power reactors with a total net capacity of some 177,000 MWe are planned and over 320 more are proposed. Energy security concerns and greenhouse constraints on coal have combined with basic economics to put nuclear power back on the agenda for projected new capacity in many countries.

2.9 Nuclear Plant Licensing

The Nuclear Regulatory Commission (NRC) is responsible for licensing and regulating the operation of commercial nuclear power plants in the United States. Currently operating nuclear power plants have been licensed under a two-step process described in Title 10 of the Code of Federal Regulations (10 CFR) under Part 50. This process requires both a construction permit and an operating license.

In an effort to improve regulatory efficiency and add greater predictability to the process, in 1989 the NRC established alternative licensing processes in 10 CFR Part 52 that included a combined license. This process combines a construction permit and an operating license with conditions for plant operation.

To construct or operate a nuclear power plant, an applicant must submit a safety analysis report. This document contains the design information and criteria for the proposed reactor and comprehensive data on the proposed site. It also discusses various hypothetical accident situations and the safety features of the plant that would prevent accidents or lessen their effects. In addition, the application must contain a comprehensive assessment of the environmental impact of the proposed plant. A prospective licensee also must submit information for antitrust reviews of the proposed plant.

When an application to construct a nuclear plant is received, the NRC staff determines whether it contains sufficient information to satisfy Commission requirements for a detailed review. If the application is accepted, the NRC holds a public meeting near the proposed site to familiarize the public with the safety and environmental aspects of the proposed application, including the planned location and type of plant, the regulatory process, and the terms for public participation in the licensing process. Numerous public meetings of this type are held during the course of the reactor licensing process.

The NRC staff then reviews the application to determine whether the plant design meets all applicable regulations (10 CFR Parts 20, 50, 73, and 100). The review includes, in part:

- characteristics of the site, including surrounding population, seismology, meteorology, geology, and hydrology;
- design of the nuclear plant;
- anticipated response of the plant to hypothetical accidents;
- plant operations including the applicant's technical qualifications to operate the plant;
- discharges from the plant into the environment (i.e., radiological effluents); and
- emergency plans.

When the NRC completes its review, it prepares a safety evaluation report summarizing the anticipated effect of the proposed facility on public health and safety.

A combined license authorizes construction of the facility much like a construction permit would under the two-step process (Part 50). It must contain essentially the same information required in an application for an operating license issued under 10 CFR Part 50 and specify the inspections, tests, and analyses that the applicant must perform. It also specifies acceptance criteria that are necessary to provide reasonable assurance that the facility has been constructed and will be operated in agreement with the license and applicable regulations. If the application does not reference an early site permit and design certification, then the NRC reviews the technical and environmental information as described for the two-step licensing process. There is also a mandatory hearing for a combined license.

2.10   Current Commercial Nuclear Plants

Pressurized Water Reactors

In a typical commercial pressurized light-water reactor, the core (1) inside the reactor vessel creates heat, (2) pressurized water in the primary coolant loop carries the heat to the steam generator, (3) inside the steam generator, heat from the steam, and (4) the steam line directs the steam to the main turbine, causing it to turn the turbine generator, which produces electricity. The unused steam is exhausted in to the condenser. The reactor's core contains fuel assemblies that are cooled by water circulated using electrically powered pumps. These pumps

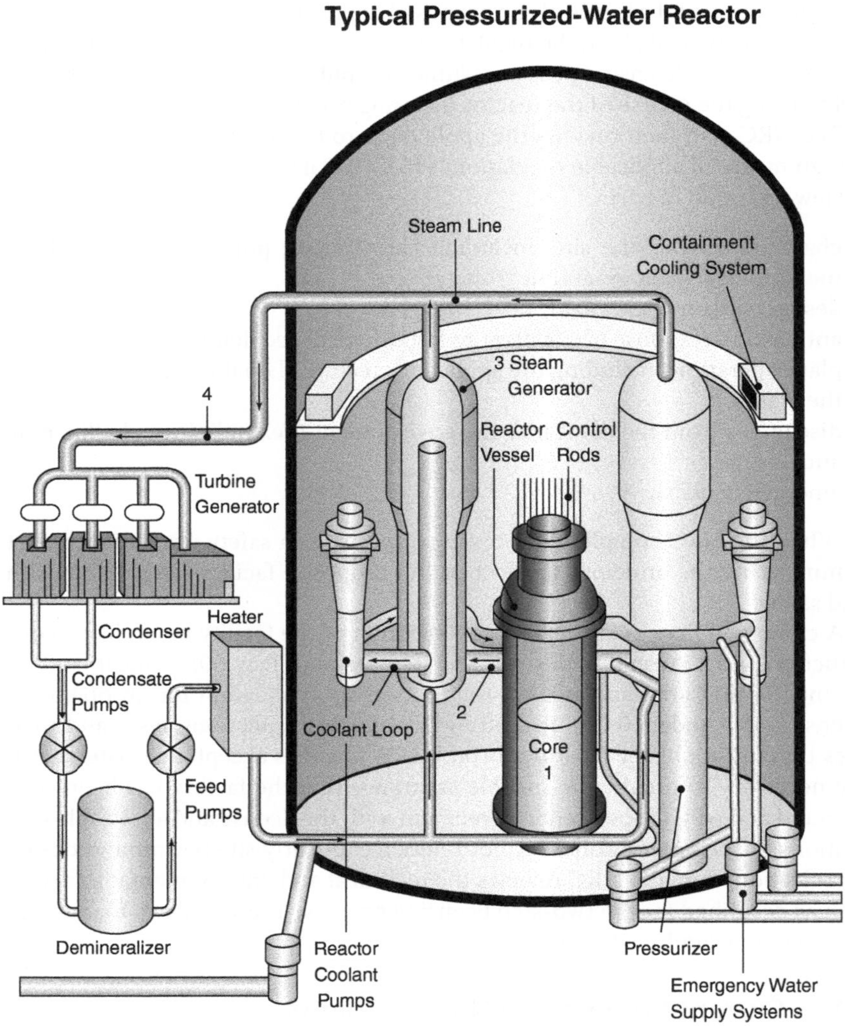

Typical Pressurized-Water Reactor

Steam Line

Containment
Cooling System

4

3 Steam
Generator

Reactor Control
Vessel Rods

Turbine
Generator

Heater

Condenser

Condensate
Pumps

Coolant Loop

2

Core
1

Feed
Pumps

Demineralizer

Reactor
Coolant
Pumps

Pressurizer

Emergency Water
Supply Systems

Figure 2.1 Typical pressurized water reactor (courtesy of U.S. NRC).

and other operating systems in the plant receive their power from the electrical grid. If off-site power is lost emergency cooling water is supplied by other pumps, which can be powered by on-site diesel generators. Other safety systems, such as the containment cooling system, also need power. PWRs contain between 150 and 200 fuel assemblies.

A typical PWR is shown in Figure 2.1.

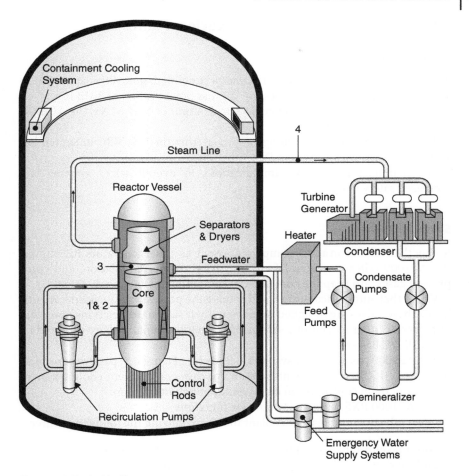

Figure 2.2 Typical boiling water reactor.

Boiling Water Reactors

In a typical commercial boiling water reactor (see Figure 2.2), the steam–water mixture leaves the top of the core and it is sent to the main turbine. Boiling water reactors contain between 370 and 800 fuel assemblies.

The reactor's core contains fuel assemblies that are cooled by water circulated using electrically powered pumps. These pumps and other operating systems in the plant receive their power from the electrical grid. If off-site power is lost, emergency cooling water is supplied by other pumps, which can be powered by on-site diesel generators. Other safety systems, such as the containment cooling system, also need electric power.

2.11   Evolutionary Commercial Nuclear Plants

Westinghouse AP1000

The AP1000 is a two-loop PWR that uses a simplified, innovative approach to safety [5]. Its main characteristics are: a gross power rating of 3415 megawatt thermal (MWt) and a nominal net electrical output of 1117 MWe, the AP1000, with a 157-fuel-assembly core.

The AP1000 received final design approval from the U.S. NRC in September 2004 and design certification in December 2005.

The manufacturer claims simplifications in overall safety systems and increased safety margins.

Safety Features

Rather than relying on active components, such as diesel generators and pumps, the AP1000 relies on natural forces—gravity, natural circulation, and compressed gases—to keep the core and the containment from overheating.

Valves that automatically align and actuate the passive safety systems are utilized. These valves are designed to move to their safeguard positions upon loss of power or upon receipt of a safeguards actuation signal and are powered by Class 1E DC batteries.

The AP1000 passive safety systems include (see Figure 2.3) passive core cooling system (PXS), containment isolation, passive containment cooling system (PCS), and main control room emergency habitability system.

The AP1000 PXS performs two major functions:

safety injection and reactor coolant makeup, utilizing the following sources:

- core makeup tanks accumulators, in-containment refueling water storage tank, and in-containment passive long-term recirculation
- passive residual heat removal, utilizing passive residual heat removal heat exchanger.

Safety injection sources are connected directly to two nozzles dedicated for this purpose on the reactor vessel. These connections have been used on two-loop plants.

The PCS consists of the following components: Air inlet and exhaust paths that are incorporated in the shield building structure. An air baffle that is located between the steel containment vessel and the concrete shield building. A passive containment cooling water storage tank that is incorporated in the shield-building structure above the containment. A water distribution system. An ancillary water storage tank and two recirculation pumps for on-site storage of additional PCS cooling water, heating to avoid freezing, and for maintaining proper water chemistry.

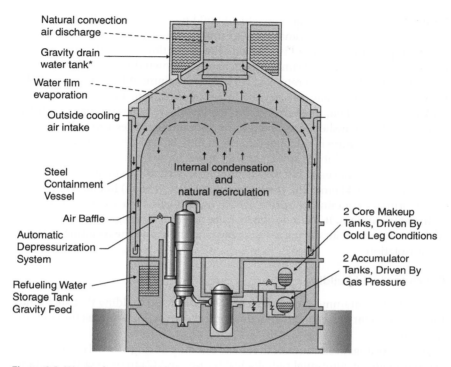

Natural convection
air discharge

Gravity drain
water tank*

Water film
evaporation

Outside cooling
air intake

Steel
Containment
Vessel

Air Baffle

Automatic
Depressurization
System

Refueling Water
Storage Tank
Gravity Feed

Internal condensation
and
natural recirculation

2 Core Makeup
Tanks, Driven By
Cold Leg Conditions

2 Accumulator
Tanks, Driven By
Gas Pressure

Figure 2.3 Westinghouse AP1000 containment and core cooling passive system (courtesy of Westinghouse).

Natural Circulation

The PCS is able to effectively cool the containment following an accident such that the design pressure is not exceeded, and the pressure is rapidly reduced. The steel containment vessel itself provides the heat transfer surface that allows heat to be removed from inside the containment and rejected to the atmosphere. Heat is removed from the containment vessel by a natural circulation flow of air through the annulus formed by the outer shield building and the steel containment vessel it houses. Outside air is pulled in through openings near the top of the shield building and pulled down, around the baffle and then flows upward out of the shield building.

The flow of air is driven by the chimney effect of air heated by the containment vessel rising and finally exhausting up through the central opening in the shield-building roof.

Water Evaporation

If needed, the air cooling can be supplemented by water evaporation on the outside of the containment shell. The water is drained by gravity from a tank located on top of the containment shield building. Three normally closed, fail-open valves will open automatically to initiate the water flow if a high containment pressure threshold is reached. The water flows from the top, outside domed surface of the steel containment shell and down the side walls, allowing heat to be transferred and removed from the containment by evaporation. The water tank has sufficient capacity for 3 days of operation, after which time the tank could be refilled, most likely from the ancillary water storage tank. If the water is not replenished after 3 days, the containment pressure will increase, but the peak pressure is calculated to reach only 90% of design pressure. After 3 days, air cooling alone is sufficient to remove decay heat.

The containment vessel is a freestanding steel structure with a wall thickness of 1.75 inches (4.44 cm). The containment is 130 ft (39.6 m) in diameter.

The primary containment prevents the uncontrolled release of radioactivity to the environment. It has a design leakage rate of 0.10 weight percent per day of the containment air during a design basis accident and the resulting containment isolation.

Concrete Shield Building

The AP1000 containment design incorporates a shield building that surrounds the containment vessel and forms the natural convection annulus for containment cooling.

The two primary functions of the shield building during normal operation are (1) to provide an additional radiological barrier and (2) to protect the containment vessel from external events, such as tornados and tornado-driven objects.

2.12 Advanced Reactors

Reactor designers are developing a number of small light-water reactor (LWR) and non-LWR designs employing innovative approaches. These designs could have applicability for location near the centers of load, or for isolated areas. Typical specifications are

Reactor power	30 MWt
Electrical output:	10 MWe
Outlet conditions	510°C
Coolant	Liquid metal (sodium)
Fuel design	18 hexagonal fuel assemblies: U-10% Zr alloy with 19.9% enrichment
Refueling	30 years

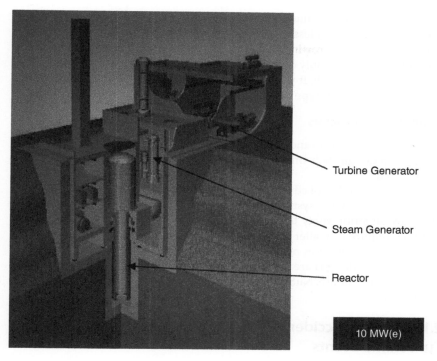

Turbine Generator

Steam Generator

Reactor

10 MW(e)

Figure 2.4 Toshiba concept for underground nuclear power plant (courtesy of Toshiba Corp.).

Underwater Nuclear Power Generating Plants

No turbines anymore (or tidal power, for that matter; see Figure 2.4). France already gets a large majority of its power from nuclear reactors, and the country clearly is not shy about continuing to innovate in the field. The state naval company, DCNS, announced plans to develop and build nuclear reactors designed to sit on the sea floor and send power back to shore.

The reactors, called Flexblue, will range from 50 to 250 MW (compare to standard large, land-based reactors, on the order of 800–1200 MW). The next phase of development will involve a DCNS collaboration with nuclear companies Areva, EDF, and others. Over 2 years, they hope to establish commercial viability as well as address safety and security concerns with the underwater concept, as well as simply ironing out the technical details.

The Flexblue reactors will come in the form of a 100-m long cylinder, with a diameter of about 15 m. They will be moored to the sea floor at depths of between 60 and 100 m, no more than a few kilometers from shore. The CEO of

DCNS suggested that siting the reactors underwater will reduce risks of proliferation and make them less vulnerable to terrorist attacks.

This idea takes the growing trend toward small modular reactors (SMRs) and adds a twist that probably does increase security, but clearly adds some technical complexity as well. It will be interesting to see if the initial studies can be completed and a prototype installed by 2016, as the company hopes.

Small Modular Reactors

SMRs range between 45 and 220 MWe, and because of their modular and easily expandable design their implementation can be very flexible based on future energy needs.

There are a number of advantages to using an SMR design. They are scalable, most modular reactor systems are factory manufactured in North America, they have superior safety features, construction time frame is much quicker, and they require a smaller up-front capital investment. SMRs also use much less space, about 20 acres per reactor.

Several SMR designs are being reviewed including designs from Babcock & Wilcox, Westinghouse, NuScale Power, and the Korea Atomic Energy Research Institute (KAERI).

2.13 Nuclear Accidents: Three Mile Island, Chernobil, and Fukushima Events

Three Mile Island

On March 28, 1979, the debate over nuclear power safety moved from the hypothetical to reality. An accident at unit 2 of the Three Mile Island plant in Pennsylvania melted about half of the reactor's core and for a time generated fear that widespread radioactive contamination would result [1]. The crisis ended without a major release of dangerous forms of radiation or a need to order a general evacuation, but it pointed out that new approaches to nuclear regulation were essential. In the aftermath of the accident, the NRC placed much greater emphasis on operator training and "human factors" in plant performance, severe accidents that could occur as a result of small equipment failures (as occurred at Three Mile Island), emergency planning, plant operating histories, and other matters.

Main electrical and instrumentation issues that were contributors to the accident:

- shortcomings in the instrumentation system design,
- poor display and annunciation of events, lack of prioritization of alarms, and
- lack of specific operator required information.

The accident began about 4:00 AM on March 28, 1979, when the plant experienced a failure in the secondary, nonnuclear section of the plant. The main feedwater pumps stopped running, caused by either a mechanical or electrical failure, which prevented the steam generators from removing heat. First the turbine, then the reactor automatically shut down. Immediately, the pressure in the primary system (the nuclear portion of the plant) began to increase. To prevent that pressure from becoming excessive, the pilot-operated relief valve (a valve located at the top of the pressurizer) opened. The valve should have closed when the pressure decreased by a certain amount, but it did not.

Signals available to the operator failed to show that the pressurizer relief valve was still open. As a result, cooling water poured out of the stuck-open valve and caused the core of the reactor to overheat (*Instrumentation Issue*).

As coolant flowed from the core through the pressurizer, the instruments available to reactor operators provided confusing information. There was no instrument that showed the level of coolant in the core. Instead, the operators judged the level of water in the core by the level in the pressurizer, and since this was high, they assumed that the core was properly covered with coolant (*Instrumentation Issue*).

In addition, there was no clear signal that the pilot-operated relief valve was open. As a result, as alarms rang and warning lights flashed, the operators did not realize that the plant was experiencing a loss-of-coolant accident (*Instrumentation Issue*).

The operators took a series of actions that made conditions worse by simply reducing the flow of coolant through the core.

Because adequate cooling was not available, the nuclear fuel overheated to the point at which the zirconium cladding (the long metal tubes which hold the nuclear fuel pellets) ruptured and the fuel pellets began to melt. It was later found that about one-half of the core melted during the early stages of the accident. Although the TMI-2 plant suffered a severe core meltdown, the most dangerous kind of nuclear power accident up to that time, it did not produce the worst-case consequences that reactor experts had long feared. In a worst-case accident, the melting of nuclear fuel would lead to a breach of the walls of the containment building and release massive quantities of radiation to the environment. But this did not occur as a result of the Three Mile Island accident.

Impact of the Accident

The accident was caused by a combination of personnel error, design deficiencies, and component failures. There is no doubt that the accident at Three Mile Island permanently changed the nuclear industry.

Following are some of the major changes which have occurred since the accident:

- Upgrading and strengthening of plant design and equipment requirements. This includes instrumentation, monitoring, and controls, operator training, fire protection, piping systems, auxiliary feedwater systems, containment building isolation, reliability of individual components (pressure relief valves and electrical circuit breakers), and the ability of plants to shut down automatically;

- Regular analysis of plant performance by senior NRC managers who identify those plants needing additional regulatory attention, expansion of NRC's resident inspector program; and
- Expansion of performance-oriented as well as safety-oriented inspections and the use of risk assessment to identify vulnerabilities of any plant to severe accidents.

Chernobyl Accident

The Chernobyl disaster was a catastrophic nuclear accident that occurred on April 26, 1986 at the Chernobyl Nuclear Power Plant in Ukraine (then officially Ukrainian SSR) [2], which was under the direct jurisdiction of the central authorities of the Soviet Union. An explosion and fire released large quantities of radioactive contamination into the atmosphere, which spread over much of Western USSR and Europe.

Main electrical and instrumentation issues that were contributors to the accident:

- Improper design of the backup diesel generators, being unable to load in the time required by the safety analysis of the plant,
- Improper testing of the generator inertia energy,
- Testing the backup generators without a complete test plan, and
- Testing the backup generators without an approved test plan.

The Chernobyl disaster is widely considered to have been one of the worst nuclear power plant accident in history; it is one of only two classified as a level 7 event on the International Nuclear Event Scale (the other being the Fukushima Daiichi nuclear disaster in 2011). The battle to contain the contamination and avert a greater catastrophe ultimately involved over 500,000 workers and cost an estimated 18 billion dollars. The official Soviet casualty count of 31 deaths could not be confirmed. Long-term effects such as cancers and deformities are still being accounted for.

The disaster began during a systems test on Saturday, April 26, 1986 at reactor number four of the Chernobyl plant, which is near the city of Prypiat and in proximity to the administrative border with Belarus and Dnieper river. There was a sudden power output surge, and when an emergency shutdown was attempted, a more extreme spike in power output occurred, which led to a reactor vessel rupture and a series of explosions. These events exposed the graphite moderator of the reactor to air, causing it to ignite. The resulting fire sent a plume of highly radioactive smoke fallout into the atmosphere and over an extensive geographical area, including Pripyat. The plume drifted over large parts of the western Soviet Union and Europe. From 1986 to 2000, 350,400 people were evacuated and resettled from the most severely contaminated areas of Belarus, Russia, and Ukraine. According to official post-Soviet data, about 60% of the fallout landed in Belarus.

The accident raised concerns about the safety of the Soviet nuclear power industry, as well as nuclear power in general, slowing its expansion for a number of years and forcing the Soviet government to become less secretive about its procedures. The government cover-up of the Chernobyl disaster was a "catalyst" for glasnost, which paved the way for reforms leading to the Soviet collapse.

Since cooling pumps require electricity to cool a reactor after a SCRAM, in the event of a power grid failure, Chernobyl's reactors had three backup diesel generators; these could start up in 15 s, but took 60–75 s to attain full speed and reach the 5.5-MW) output required to run one main pump. Thus, there was a 1-min gap in providing cooling to the reactor and this was considered an unacceptable safety risk. It was theorized at the time that the remaining rotational inertia of the steam turbine generator after a trip, could be utilized to generate the required electrical power to power the safety systems until the backup diesel generators were online. Analysis indicated the energy to be sufficient to run the coolant pumps for 45 s.

However, this capability still needed to be confirmed experimentally, and previous tests had ended unsuccessfully. An initial test carried out in 1982 showed that the excitation voltage of the turbine generator was insufficient maintain the desired magnetic field after the turbine trip. The system was modified, and the test was repeated in 1984 but again proved unsuccessful. In 1985, the tests were attempted a third time but also yielded negative results. The test procedure was to be repeated again in 1986, and it was scheduled to take place during the maintenance shutdown of reactor 4.

According to the test parameters, the thermal output of the reactor should have been no lower than 700 MW at the start of the experiment, but the test procedure was not officially approved.

The entire Chernobyl power plant site was eventually shut down on December 15, 2000.

Containment of the Reactor

The Chernobyl reactor is now enclosed in a large concrete sarcophagus, which was built quickly to allow continuing operation of the other reactors at the plant.

A New Safe Confinement was to have been built by the end of 2005; however, it has suffered ongoing delays and as of 2010, when construction finally began, it is now expected to be completed in 2018. The structure is being built adjacent to the existing shelter and will be slid into place on rails. It is to be a metal arch 105 m (344 ft) high and spanning 257 m (843 ft), to cover both unit 4 and the hastily built 1986 structure. The Chernobyl Shelter Fund, set up in 1997, has received €810 million from international donors and projects to cover this project and previous work. It and the Nuclear Safety Account, also applied to Chernobyl decommissioning, are managed by the European Bank for Reconstruction and Development.

By 2002, roughly 15,000 Ukrainian workers were still working within the Zone of Exclusion, maintaining the plant and performing other containment- and research-related tasks, often in dangerous conditions. A handful of Ukrainian scientists work inside the sarcophagus, but outsiders are rarely granted access. In 2006, an Australian 60 Minutes team led by reporter Richard Carleton and producer Stephen Rice were allowed to enter the sarcophagus for 15 min and film inside the control room.

Fukushima Accident – 2011

The Fukushima I Nuclear Power Plant comprised six separate boiling water reactors originally designed by General Electric (GE) and maintained by the Tokyo Electric Power Company (TEPCO). At the time of the Tōhoku earthquake on March 11, 2011, reactors 4–6 were shut down in preparation for refueling. However, their spent fuel pools still required cooling. Units 1–3 automatically shut down in accordance with the signal, indicating "increasing seismic acceleration" [3].

Following the reactor and generator shutdown, functionality of any external power source system equipment was also lost. However, ac power for each unit was supplied by emergency diesel generators.

The largest tsunami wave was 13 m high and hit 50 min after the initial earthquake, overwhelming the plant's seawall, which was 10 m high. Subsequently, due to the tsunami which struck the station all ac power sources at units 1–5 were lost, except for one air-cooled unit at unit 6.

Furthermore, DC power sources were also lost due to inundation at units 1, 2, and 4, and such function was also lost at unit 3 as a result of batteries being depleted, resulting in a situation where all power being supplied to units 1–4 was lost.

As a result of the loss of ac power sources, existing cooling function for reactors and spent fuel pools was lost, and cooling was attempted using temporary power sources and alternative coolant injection, but the situation escalated into one where fuel in the core was damaged and radioactive material released into the environment.

Summary of the Accident
- Following a major earthquake, a 15-m tsunami disabled the power supply and cooling of three Fukushima Daiichi reactors, causing a nuclear accident on March 11, 2011.
- All three cores largely melted in the first 3 days.

- The accident was rated 7 on the scale, due to high radioactive releases in the first few days. Four reactors are written off—2719 MWe net.
- After 2 weeks, the three reactors (units 1–3) were stable with water addition but no proper heat sink for removal of decay heat from fuel. By July, they were being cooled with recycled water from the new treatment plant. Reactor temperatures had fallen to below 80°C at the end of October, and official "cold shutdown condition" was announced in mid-December 2011.
- Apart from cooling, the basic ongoing task was to prevent release of radioactive materials, particularly in contaminated water leaked from the three units.
- There have been no deaths or cases of radiation sickness from the nuclear accident, but over 100,000 people had to be evacuated from their homes to ensure this. Government nervousness has delayed their return.

The Great East Japan Earthquake of magnitude 9.0 at 2.46 PM on Friday March 11, 2011 did considerable damage in the region, and the large tsunami it created caused very much more. The earthquake was centered 130 km offshore the city of Sendai in Miyagi prefecture on the eastern cost of Honshu Island (the main part of Japan) and was a rare and complex double quake giving a severe duration of about 3 min. Japan moved a few meters east, and the local coastline subsided half a meter. The tsunami inundated about 560 km^2 and resulted in a human death toll of over 19,000 and much damage to coastal ports and towns with over a million buildings destroyed or partly collapsed. Eleven reactors at four nuclear power plants in the region were operating at the time, and all shut down automatically when the quake hit. Subsequent inspection showed no significant damage to any form from the earthquake. The operating units which shut down were TEPCO's Fukushima Daiichi 1–3, and Fukushima Daini 1–4, Tohoku's Onagawa 1–3, and Japco's Tokai, total 9377 MWe net. Fukushima Daiichi units 4–6 were not operating at the time, but were affected. The main problem initially centered on Fukushima Daiichi units 1–3. Unit 4 became a problem on day five. The reactors proved robust seismically, but vulnerable to the tsunami. Power, from grid or backup generators, was available to run the residual heat removal (RHR) system cooling pumps at eight of the eleven units, and despite some problems they achieved "cold shutdown" within about 4 days. The other three, at Fukushima Daiichi, lost power at 3.42 PM, almost 1 h after the quake, when the entire site was flooded by the 15-m tsunami. This disabled 12 of 13 backup generators on site and also the heat exchangers for dumping reactor waste heat and decay heat to the sea. The three units lost the ability to maintain proper reactor cooling and water circulation functions. Electrical switchgear was also disabled. Thereafter, many weeks of focused work centered on restoring heat removal from the reactors and coping with overheated spent fuel ponds. This was undertaken by hundreds of TEPCO employees as well as some contractors, supported by firefighting and military personnel. Some of the TEPCO staff had lost homes, and even families, in the tsunami, and were initially living in temporary accommodation under great difficulties and privation, with some personal risk. A hardened emergency response center on site was unable to be used in grappling with the situation due to radioactive contamination. Three TEPCO employees at the Daiichi and Daini plants were killed directly by the earthquake and tsunami, but there have been no fatalities from the nuclear accident. Among hundreds of aftershocks, an earthquake with magnitude 7.1, closer to Fukushima than the 11 March one, was experienced on 7 April, but without further damage to the plant. On 11 April, a magnitude 7.1 earthquake and on 12 April a magnitude 6.3 earthquake, both with epicenter at Fukushima-Hamadori, caused no further problems.

The Two Fukushima Plants and Their Siting

The Daiichi (first) and Daini (second) Fukushima plants are sited about 11 km apart on the coast, Daini to the south. The recorded seismic data for both plants—some 180 km from the epicenter—shows that 550 Gal (0.56 g) was the maximum ground acceleration for Daiichi, and 254 Gal was maximum for Daini. Daiichi units 2, 3, and 5 exceeded their maximum response acceleration design basis in the E–W direction by about 20%. The recording was over 130–150 s. (All nuclear plants in Japan are built on rock—ground acceleration was around 2000 Gal a few kilometers north, on sediments.) The original design basis tsunami height was 3.1 m for Daiichi based on assessment of the 1960 Chile tsunami, and so the plant had been built about 10 m above sea level with the seawater pumps 4 m above sea level. The Daini plant as built 13 m above sea level. In 2002, the design basis was revised to 5.7 m above, and the seawater pumps were sealed. Tsunami heights coming ashore were about 15 m, and the Daiichi turbine halls were under some 5 m of seawater until levels subsided. Daini was less affected. The maximum amplitude of this tsunami was 23 m at point of origin, about 180 km from Fukushima.

In the past century, there have been eight tsunamis in the region with maximum amplitudes at origin above 10 m (some much more), these having arisen from earthquakes of magnitude 7.7–8.4, on average one every 12 years. Those in 1983 and in 1993 were the most recent affecting Japan, with maximum heights at origin of 14.5 and 31 m, respectively, both induced by magnitude 7.7 earthquakes. The June 1896 earthquake of estimated magnitude 7.6 produced a tsunami with run-up height of 38 m in the Tohoku region, killing 27,000 people. The tsunami countermeasures taken when Fukushima Daiichi was designed and sited in the 1960s were considered acceptable in relation to the scientific knowledge then, with low recorded run-up heights for that particular coastline. But through to the 2011 disaster, new scientific knowledge emerged about the likelihood of a large earthquake and resulting major tsunami. However, this did not lead to any major action by either the plant operator, TEPCO, or government regulators, notably the Nuclear & Industrial Safety Agency (NISA).

The tsunami countermeasures could also have been reviewed in accordance with IAEA guidelines which required taking into account high tsunami levels, but NISA continued to allow the Fukushima plant to operate without sufficient countermeasures, despite clear warnings. A report from the Japanese government's Earthquake Research Committee on earthquakes and tsunamis off the Pacific coastline of northeastern Japan in February 2011 was due for release in April, and might have brought about changes. The document includes analysis of a magnitude 8.3 earthquake that is known to have struck the region more than 1140 years ago, triggering enormous tsunamis that flooded vast areas of Miyagi and Fukushima prefectures. The report concludes that the region should be alerted of the risk of a similar disaster striking again. The 11 March earthquake measured magnitude 9.0 and involved substantial shifting of multiple sections of seabed over a source area of 200×400 km. Tsunami waves devastated wide areas of Miyagi, Iwate, and Fukushima prefectures.

Events at Fukushima Daiichi 1–3 and 4

It appears that no serious damage was done to the reactors by the earthquake, and the operating units 1–3 were automatically shut down in response to it, as designed. At the same time, all six external power supply sources were lost due to earthquake damage, so the emergency diesel generators located in the basements of the turbine buildings started up. Initially cooling would have been maintained through the main steam circuit bypassing the turbine and going through the condensers. Then 41 min later, the first tsunami wave hit, followed by a second 8 min later. These submerged and damaged the seawater pumps for both the main condenser circuits and the auxiliary cooling circuits, notably the RHR cooling system. They also drowned the diesel generators and inundated the electrical switchgear and batteries, all located in the basements of the turbine buildings (the one surviving air-cooled generator was serving units 5 and 6). The only surviving emergency generator was located on the first floor of a building near unit 6. Reactor 6 and its sister unit, reactor 5, would weather the crisis without serious damage, thanks in part to that generator. There was a station blackout, and the reactors were isolated from their ultimate heat sink. The tsunamis also damaged and obstructed roads, making outside access difficult. All this put those reactors 1–3 in a dire situation and led the authorities to order, and subsequently extend, an evacuation while engineers worked to restore power and cooling. The 125-V DC batteries for units 1 and 2 were flooded and failed, leaving them without instrumentation, control, or lighting. Unit 3 had battery power for about 30 h. At 7.03 PM, Friday 11 March a nuclear emergency was declared, and at 8.50 PM the Fukushima Prefecture issued an evacuation order for people within 2 km of the plant. At 9.23 PM, the Prime Minister extended this to 3 km, and at 5.44 AM on 12 March he extended it to 10 km. He visited the plant soon after. On Saturday 12 March, he extended the evacuation zone to 20 km.

Inside the Fukushima Daiichi Reactors

The Fukushima Daiichi reactors are GE boiling water reactors of an early (1960s) design supplied by GE, Toshiba, and Hitachi, with what is known as a Mark I containment. Reactors 1–3 came into commercial operation in 1971–1975. Reactor power is 460 MWe for unit 1, 784 MWe for units 2–5, and 1100 MWe for unit 6.

Summary: Major fuel melting occurred early on in all three units, though the fuel remains essentially contained except for some volatile fission products vented early on, or released from unit 2 in mid-March, and some soluble ones which were leaking with the water, especially from unit 2, where the containment is evidently breached. Cooling still needs to be provided from external sources, now using treated recycled water, while work continues to establish a stable heat removal path from the actual reactors to external heat sinks. Temperatures at the bottom of the reactor pressure vessels have decreased to well below boiling point and are stable. Access has been gained to all three reactor buildings, but dose rates remain high inside. Nitrogen is being injected into all three containment vessels and pressure vessels. TEPCO declared "cold shutdown condition" in mid-December when radioactive releases had reduced to minimal levels.

Fuel Ponds: Developing Problems

Used fuel needs to be cooled and shielded. This is initially by water, in ponds. After about 3 years under water, used fuel can be transferred to dry storage, with air ventilation simply by convection. Used fuel generates heat, so the water is circulated by electric pumps through external heat exchangers, so that the heat is dumped and a low temperature is maintained. There are fuel ponds near the top of all six reactor buildings at the Daiichi plant, adjacent to the top of each reactor so that the fuel can be unloaded under water when the top is off the reactor pressure vessel and it is flooded. The ponds hold some fresh fuel and some used fuel, pending its transfer to the central used/spent fuel storage on site. (There is some dry storage on site to extend the plant's capacity.)

At the time of the accident, in addition to a large number of used fuel assemblies, unit 4's pond also held a full core load of 548 fuel assemblies while the reactor was undergoing maintenance, these having been removed at the end of November.

A separate set of problems arose as the fuel ponds, holding fresh and used fuel in the upper part of the reactor structures, were found to be depleted in water. The primary cause of the low water levels was loss of cooling circulation to external heat exchangers, leading to elevated temperatures and probably boiling, especially in heavily loaded unit 4. Here the fuel would have been uncovered in about 7 days due to water boiling off. However, the fact that unit 4 was unloaded meant that there was a large inventory of water at the top of the structure, and enough of this replenished the fuel pond to prevent the fuel becoming uncovered—the minimum level reached was about 1.2 m above the fuel on about 22 April.

After the hydrogen explosion in unit 4 early on Tuesday 15 March, TEPCO was told to implement injection of water to unit 4 pond which had a particularly high heat load (3 MW) from 1331 used fuel assemblies in it, so it was the main focus of concern. It needed the addition of about 100 m³/day to replenish it after circulation ceased.

From Tuesday 15 March, attention was given to replenishing the water in the ponds of units 1–3 as well. Initially, this was attempted with fire pumps but from 22 March a concrete pump with 58-m boom enabled more precise targeting of water through the damaged walls of the service floors. There was some use of built-in plumbing for unit 2. Analysis of radionuclides in water from the used fuel ponds suggested that some of the fuel assemblies might be damaged, but the majority were intact.

There was concern about structural strength of unit 4 building, so support for the pond was reinforced by the end of July.

New cooling circuits with heat exchangers adjacent to the reactor buildings for all four ponds were commissioned after a few months, and each reduced the pool temperature from 70°C to normal in a few days. Each has a primary circuit within the reactor and waste treatment buildings and a secondary circuit dumping heat through a small dry cooling tower outside the building.

The next task was to remove the salt from those ponds which had seawater added, to reduce the potential for corrosion.

In July 2012, the first of the 204 fresh fuel assemblies were removed for the unit 4 pond and transferred to the central spent fuel pool for detailed inspection to check damage, particularly corrosion. They were found to have no deformation or corrosion. These comprise 783 spent fuel plus the full fuel load of 548.

The central spent fuel pool on site holds about 60% of the Daiichi used fuel and is immediately west (inland) of unit 4. It lost circulation with the power outage, and temperature increased to 73°C by the time mains power and cooling were restored after 2 weeks.

Summary: The new cooling circuits with external heat exchangers for the four ponds are working well. Temperatures are normal. Analysis of water confirmed that most fuel rods are intact.

There are no clear plans for decommissioning the plant, but the plant management estimate it will take 30 or 40 years.

A frozen soil barrier has been constructed in an attempt to prevent further contamination of seeping groundwater by melted-down nuclear fuel, but in July 2016 TEPCO revealed that the ice wall had failed to totally stop groundwater from flowing in and mixing with highly radioactive water inside the wrecked reactor buildings, adding that they are "technically incapable of blocking off groundwater with the frozen wall."

In February 2017, TEPCO released images taken inside reactor 2 by a remote-controlled camera, showing that there is a 2-m (6.5 ft) wide hole in the metal grating under the pressure vessel in the reactor's primary containment vessel, which could have been caused by fuel escaping the pressure vessel, indicating a meltdown/melt-through had occurred, through this layer of containment. Radiation levels of about 210 Sv/h were subsequently detected inside the unit 2 containment vessel. These values are in the context of undamaged spent fuel which has typical values of 270 Sv/h, after 10-years of cold shutdown, with no shielding.

Questions and Problems

2.1 Provide the names and the functions of power systems which are automatically actuated under loss of off-site power (loss of the grid power).

2.2 For the Three Mile Island Accident:
a) Name the main causes for the accident,
b) Describe the contribution of electrical issues to this accident.
c) What measures were instituted to prevent a similar event?
d) What is the present condition of the plant?

2.3 For the Chernobil accident:
a) Name the main causes for the accident.
b) Name the contribution of electrical issues to this accident (the emergency diesel generators are considered part of the electrical equipment at the plant).
c) What measures were instituted to prevent a similar event?
d) What is the present condition of the plant?

2.4 Fukushima accident
a) Name the main causes for the accident at Fukushima.
b) Name the contribution of electrical issues to this accident (the emergency diesel generators are considered part of the electrical equipment at the plant).
c) What measures were instituted to prevent a similar event?
d) What is the present condition of the plant?

2.5 Increased generating capacity
Increased nuclear capacity in some countries is resulting from the uprating of existing plants. This is a highly cost-effective way of bringing on new capacity.

Provide a summary of the main electrical systems/components that would need to be reviewed as related to increased power output of the nuclear plant (refer to the plant main one-line diagram, Figure 1.1)

References

1 US NRC, "Lessons learned from the Three Mile Island - Unit 2 Advisory Panel," NUREG/CR-6252.
2 NUREG 1250 Report on the Accident at the Chernobyl Nuclear Power Station, January 1987.
3 Fukushima accident, USNRC report, ML 1605/ML 16054A139.
4 Key World Energy Statistics 2009.
5 Westinghouse Electric Corporation AP1000 PWR Design Aspects.

3

Special Regulations and Requirements

3.1 Regulations

This section lists regulations that must be followed and which apply to electrical systems for nuclear power plants.

10 CFR Part 50, "Domestic Licensing of Production and Utilization Facilities" [2]

Section 10: Energy, Part 50: domestic licensing of production and utilization facilities.

10 CFR Part 50.71(e) requires that a safety analysis report (SAR) be prepared and submitted for approval by the Nuclear Regulatory Commission (NRC). Particularly important for electrical systems is Section 8.0—Electric Power, which comprises a description of the plant connection to the grid, the plant alternating current (ac) and direct current (DC) systems.

10CFR50 APPENDIX A, General Design Criteria, Criterion 20, requires that the protection system be designed to initiate the operation of systems and components important to safety.

10CFR50 APPENDIX A, General Design Criteria, Criterion 21, requires that the protection system be designed to permit periodic testing of its functioning when the reactor is in operation. In current designs, the ability of the protection system to initiate the operation of safety systems depends on the proper performance of actuation devices; therefore, these devices are to be tested.

10CFR50 APPENDIX A, General Design Criteria, Criterion 17—Electric power systems

An on-site electric power system and an off-site electric power system should be provided to permit functioning of structures, systems, and components important to safety. The safety function for each system (assuming the other system is not functioning) should be to provide sufficient capacity and capability to assure that (1) specified acceptable fuel design limits and design conditions of the reactor coolant pressure boundary are not exceeded as a result of

Electrical Systems for Nuclear Power Plants, First Edition. Omar S. Mazzoni.
© 2019 by The Institute of Electrical and Electronic Engineers, Inc. Published 2019 by John Wiley & Sons, Inc.

anticipated operational occurrences and (2) the core is cooled and containment integrity and other vital functions are maintained in the event of postulated accidents.

The on-site electric power supplies, including the batteries, and the on-site electric distribution system should have sufficient independence, redundancy, and testability to perform their safety functions assuming a single failure.

Electric power from the transmission network to the on-site electric distribution system should be supplied by two physically independent circuits (not necessarily on separate rights of way) designed and located so as to minimize to the extent practical the likelihood of their simultaneous failure under operating and postulated accident and environmental conditions. A switchyard common to both circuits is acceptable. Each of these circuits should be designed to be available in sufficient time following a loss of all on-site ac power supplies and the other off-site electric power circuit, to assure that specified acceptable fuel design limits and design conditions of the reactor coolant pressure boundary are not exceeded. One of these circuits should be designed to be available within a few seconds following a loss-of-coolant accident to assure that core cooling, containment integrity, and other vital safety functions are maintained.

Provisions should be included to minimize the probability of losing electric power from any of the remaining supplies as a result of, or coincident with, the loss of power generated by the nuclear power unit, the loss of power from the transmission network, or the loss of power from the on-site electric power supplies.

10CFR50 APPENDIX A, General Design Criteria, Criterion 18— Inspection and Testing of Electric Power Systems.

Electric power systems important to safety should be designed to permit appropriate periodic inspection and testing of important areas and features, such as wiring, insulation, connections, and switchboards, to assess the continuity of the systems and the condition of their components. The systems should be designed with a capability to test periodically (1) the operability and functional performance of the components of the systems, such as on-site power sources, relays, switches, and buses and (2) the operability of the systems as a whole and, under conditions as close to design as practical, the full operation sequence that brings the systems into operation, including operation of applicable portions of the protection system, and the transfer of power among the nuclear power unit, the off-site power system, and the on-site power system.

10CFR50 APPENDIX A, General Design Criteria, Criterion 19—Control Room. A control room should be provided from which actions can be taken to operate the nuclear power unit safely under normal conditions and to maintain it in a safe condition under accident conditions, including loss-of-coolant accidents. Adequate radiation protection should be provided to permit access and occupancy of the control room under accident conditions without personnel receiving radiation exposures in excess of 5 rem whole body, or its equivalent to any part of the body, for the duration of the accident. Equipment at appropriate locations outside the control room should be provided (1) with a design capability for prompt hot shutdown of the reactor, including necessary instrumentation and controls to maintain the unit in a safe condition during hot shutdown and (2) with a potential capability for subsequent cold shutdown of the reactor through the use of suitable procedures.

Applicants for and holders of construction permits and operating licenses under this part who apply on or after January 10, 1997, applicants for design approvals or certifications under part 52 of this chapter who apply on or after January 10, 1997, applicants for and holders of combined licenses or manufacturing licenses under part 52 of this chapter who do not reference a standard design approval or certification, or holders of operating licenses using an alternative source term under § 50.67, should meet the requirements of this criterion, except that with regard to control room access and occupancy, adequate radiation protection should be provided to ensure that radiation exposures should not exceed 0.05 Sv (5 rem) total effective dose equivalent as defined in § 50.2 for the duration of the accident.

10 CFR 50, Appendix A, General Design Criteria, Criterion 20, requires that the protection system be designed to initiate the operation of systems and components important to safety.

10CFR50 APPENDIX A, General Design Criteria, Criterion 21, requires that the protection system be designed to permit periodic testing of its functioning when the reactor is in operation. In current designs, the ability of the protection system to initiate the operation of safety systems depends on the proper performance of actuation devices; therefore, these devices are to be tested. This safety guide describes acceptable methods of including the actuation devices in the periodic tests of the protection system during reactor operation. It does not address the frequency of such testing.

10 CFR 50 APPENDIX B, Quality Assurance Criteria, deals with notices of nonconformance and notices of violation

 I. Organization,
 II. Quality assurance program,
 III. Design control,
 IV. Procurement document control,
 V. Instructions, procedures, and drawings,
 VI. Document Control VII: Control of purchased material, equipment, and services,
 VII. Identification and control of materials, parts, and components,
 VIII. Control of special processes,
 IX. Inspection,
 X. Test control,
 XI. Control of measuring and test equipment,

XII. Handling, storage, and shipping,
XIII. Inspection, test, and operating status,
XIV. Nonconforming materials, parts, or components,
XV. Corrective action,
XVI. Quality assurance records, and
XVII. Audits.

Criterion III, "Design Control," and Criterion XI, "Test Control," of Appendix B, "Quality Assurance Criteria for Nuclear Power Plants and Fuel Reprocessing Plants," to 10 CFR Part 50 require that (1) measures be provided for verifying or checking the adequacy of design through design reviews, the use of alternative or simplified calculational methods, or the performance of a suitable testing program and (2) a test program be established to ensure that systems and components perform satisfactorily and that the test program include operational tests during nuclear power plant operation.

10 CFR 50.63, "Loss of All Alternating Current Power," requires that each light-water-cooled nuclear power plant must be able to withstand and recover from a station blackout (i.e., loss of off-site and on-site emergency ac power systems) for a specified duration. The reliability of on-site ac power sources is one of the main factors contributing to the risk of core melt as a result of a station blackout.

10 CFR - Part 21, Notice of Violations

These regulations cover all applicable Quality Assurance (QA) Inspection for new reactor licensing and vendor QA inspection reports that have either a Notice of Nonconformance or Notice of Violation within a specific criterion of 10 CFR 50 Appendix B or 10 CFR Part 21 related issue.

NRC Standard Review Plan (NUREG-0800 Standard Review Plan (SRP) for the Review of SAR for nuclear power plants)

The SRP provides guidance to US NRC staff in performing safety reviews of construction permit or operating license applications (including requests for amendments) under 10 CFR Part 50 and early site permit, design certification, combined license, standard design approval, or manufacturing license applications under 10 CFR Part 52 (including requests for amendments).

The principal purpose of the SRP is to assure the quality and uniformity of staff safety reviews. It is also the intent of this plan to make information about regulatory matters widely available and to improve communication between the NRC, interested members of the public, and the nuclear power industry, thereby increasing understanding of the NRC's review process.

Because the SRP generally describes an acceptable means of meeting the regulations, but not necessarily the only means, applications may deviate from the acceptance criteria in the SRP.

Of particular interest for this course is Chapter 8—Electric Power, which consists of the following:

8.1, Electric Power - Introduction
8.2, Offsite Power System
8.3.1, AC Power Systems (Onsite)
8.3.2, DC Power Systems (Onsite)
8.4, Station Blackout
NRC Branch Technical Positions
Branch Technical Position 8-1, Requirements on Motor-Operated Valves in the Emergency Core Cooling System Accumulator Lines
Branch Technical Position 8-2, Use of Diesel-Generator Sets for Peaking
Branch Technical Position 8-3, Rev. 3 Stability of Offsite Power Systems
Branch Technical Position 8-4, Rev. 3 Application of the Single Failure Criterion to Manually Controlled Electrically Operator Valves
Branch Technical Position 8-5, Rev. 3 Supplemental Guidance for Bypass and Inoperable Status Indication for Engineered Safety Features Systems
Branch Technical Position 8-6, Rev. 3 Adequacy of Station Electric Distribution System Voltages
Branch Technical Position 8-7, Rev. 3, Criteria for Alarms and Indications Associated with Diesel-Generator Unit Bypassed and Inoperable Status

3.2 IEEE Standards

IEEE has issued multiple standards, which are referenced in the applicable sections. The standards provide acceptable ways and methods for the design and specification of equipment for nuclear power plants.

3.3 NRC Regulatory Guides

This section lists regulatory guides that provide ways and methods to comply with the regulations. As such they are not required to be followed if another method is proven at least equally efficient and proposed for compliance.

The US NRC issues regulatory guides to describe and make available to the public methods that the NRC staff considers acceptable for use in implementing specific parts of the agency's regulations, techniques that the staff uses in evaluating specific problems or postulated accidents, and data that the staff need in reviewing applications for permits and licenses. Regulatory guides are not substitutes for regulations, and compliance with them is not required.

Some of the regulatory guides, which are particularly important to electrical systems, are as follows:

Box 3.1 Regulatory Guide 1.9 [1]

Application and Testing of Safety-Related Diesel Generators in Nuclear Power Plants

(Draft was issued as DG-1172, dated November 2006)
   This regulatory Guide provides supplemental guidance and endorses (IEEE) Standard 387, "IEEE Standard Criteria for Diesel-Generator Units Applied as Standby Power Supplies for Nuclear Power Generating Stations" (IEEE Std 387). This subject is discussed in detail under Chapter 5 of this book.

Box 3.2 Regulatory Guide 1.22

Periodic Testing of Protection System Actuation Functions

This safety guide describes acceptable methods of including the actuation devices in the periodic tests of the protection system during reactor operation. It does not address the frequency of such testing
   This guide provides general methods and provisions for compliance with General Design Criterion 20 of Appendix A to 10 CFR Part 50, "General Design Criteria for Nuclear Power Plants," which requires that the protection system be designed to initiate the operation of systems and components important to safety. General Design Criterion 21 requires that the protection system be designed to permit periodic testing of its functioning when the reactor is in operation. In current designs, the ability of the protection system to initiate the operation of safety systems depends on the proper performance of actuation devices; therefore, these devices are to be tested.

Box 3.3 Regulatory Guide 1.32

Criteria for Power Systems for Nuclear Power Plants

This guide provides general criteria and endorses IEEE Std 308 (Criteria for Class 1E Power Systems for Nuclear Power Generating Stations), with the addition of supplementary requirements.

Box 3.4   Regulatory Guide 1.41

Preoperational Testing of Redundant On-Site Electric Power Systems to Verify Proper Load Group Assignments

This guide provides requirements for the independence among redundant on-site power sources, and their load groups should be such that the successful operation of any power source and its load group is in no way affected by the partial or complete failure of any other power source and its load group.

Box 3.5   Regulatory Guide 1.47

Bypassed and Inoperable Status Indication for Nuclear Power Plant Safety Systems

This guide describes an acceptable method of complying with the requirements of IEEE Std 279[1] (Criteria for Protection Systems for Nuclear Power Generating Stations) and Appendix B to 10 CFR Part 50 with regard to indicating the inoperable status of a portion of the protection system (as defined in IEEE Std 279), systems actuated or controlled by the protection system, and auxiliary or supporting systems that must be operable for the protection system and the systems it actuates to perform their safety-related functions.

Criterion XIV, "Inspection, Test, and Operating Status," of Appendix B to 10 CFR Part 50, "Quality Assurance Criteria for Nuclear Power Plants and Fuel Reprocessing Plants," requires that measures be established for indicating the operating status of structures, systems, and components of the nuclear power plant, such as by tagging valves and switches, to prevent inadvertent operation. Section 50.55a, "Codes and Standards," of 10 CFR Part 50, requires in Paragraph (h) that protection systems meet the requirements set forth in the Institute of Electrical and Electronics Engineers "Criteria for Nuclear Power Plant Protection Systems" (IEEE Std 279). Section 4.13 of IEEE Std 279, "Criteria for Protection Systems for Nuclear Power Generating Stations," (also designated ANSI N42.7) requires that, if the protective action of some part of the protection system has been bypassed or deliberately rendered inoperative for any purpose, this fact should be continuously indicated in the control room. Applicable NRC position is provided in Regulatory Guide 1.47

---

1   IEEE 279-1971 (reaff 1978) standard is available from IEEE.

Questions and Problems

3.1 Provide the main difference between Federal Regulations and NRC Regulatory Guides.

3.2 Provide a summary description of the safety analysis report.

3.3 Name the legal regulation that requires that an on-site electric power system and an off-site electric power system be provided for the functioning of structures, systems, and components important to safety.

3.4 Describe the safety classifications of the on-site and the off-site electric power systems.

3.5 Describe the attributes of the on-site electric system.

3.6 Describe the attributes required from the off-site electric system.

3.7 Describe the plant requirements relative to the probability of losing electric power.

3.8 Name the legal regulation that requires that an on-site electric power system and an off-site electric power system be provided with capability for inspection and testing.

3.9 Name the legal regulation for control room design.

3.10 Name the legal regulation for quality assurance.

3.11 Describe the requirement for "design control."

3.12 Describe the requirement for "loss of all alternating current power."

3.13 Describe the Nuclear Regulatory Commission Standard Review Plan.

3.14 Describe the sections of the Standard Review Plan that are most important for this book.

3.15 Describe important Nuclear Regulatory Commission Branch Technical Positions as related to this book.

3.16 Describe Nuclear Regulatory Commission Regulatory Guides which are particularly important to this book.

References

1 US NRC Nuclear Regulatory Guide 1.9. "Application and Testing of Safety-Related Diesel Generators in Nuclear Power Plants."

2 10 CFR Part 50, "Domestic Licensing of Production and Utilization Facilities."

4

Unique Requirements: Class 1E Power System

This chapter offers a bridge to subsequent chapters dealing with Class 1E Electrical Systems starting with a general description followed a brief introduction to specific requirements for Class 1E equipment:

- alternating current (ac) power and distribution systems,
- direct current (DC) power and distribution systems,
- Instrumentation and control systems,
- Containment electrical penetration assemblies, and
- Emergency on-site ac power sources.

4.1 Class 1E Electrical Systems: General Description

This section provides general description of description of the Class 1E Electrical Systems and their interfaces with the nuclear plant and with other systems. Detailed descriptions are provided in Chapters 5–7. Unique requirements are discussed in references [1], [4], and [5]. The Class 1E Electrical Systems are

- Class 1E ac Power and Distribution Systems,
- Class 1E DC Power and Distribution Systems,
- Class 1E Instrumentation and Control Systems, and
- Class 1E Containment Electrical Penetration Assemblies.

In general, the specific design details are provided in a "design basis document" (DBD), which details the design fundamentals and all associated requirements. The distribution system includes all equipment in the distribution circuit from its supply circuit breaker(s) or fuses to the load terminals.

The DBD should be prepared for all of the components of the Class 1E system. As a minimum, it includes the following:

- Description of all plant events requiring emergency and normal operation of the Class 1E power systems.
- Actuation signals for operation of the Class 1E power systems.

Electrical Systems for Nuclear Power Plants, First Edition. Omar S. Mazzoni.

- All the loads connected to the Class 1E busses.
- Sequence for start-up and loading of the Class 1E power sources, including voltage, current, time, speed, and other parameters applicable to the standby power supplies, during and after the sequence of events for the plant normal and emergency conditions.
- All predictable malfunctions, accidents, environmental events, and operating modes that could potentially degrade Class 1E power systems or its projected performance, for which provisions should be incorporated.
- The limits of acceptability of transient and steady-state conditions, including environmental and operating parameters.

All normal and possible abnormal natural occurring events should be listed, including

- earthquake,
- rain,
- ice and snow,
- wind,
- floods,
- hurricane,
- lightning,
- tornado,
- extreme temperature conditions, and
- tsunami.

Postulated events should be considered, which are as follows:

- postulated accident environment (humidity, temperature, pressure, chemical properties, radiation),
- fires,
- accident and nonaccident-generated missiles, including pipe whip,
- fire protection system operation,
- accident and nonaccident-generated flooding, sprays, or jets,
- postulated loss of the preferred power supply combined with any other postulated event,
- postulated loss or failure of the preferred power supply,
- postulated loss of all alternating current electric power (station blackout),
- any single equipment or single component malfunction,
- single act, event, component failure, or circuit fault that can cause multiple equipment malfunctions, and
- equipment maintenance outage, while the plant is running.

Corrective measures should be implemented for any identified possible events. The following main requirements apply to all Class 1E equipment:

1. Redundancy. The Class 1E electric loads need to be separated into two or more redundant load groups., which should be independent.
2. The protective actions of each load group should be independent of the protective actions provided by redundant load groups.
3. Two or more load groups may have a common power supply if the consequences of the loss of the common power supply to the load groups under design basis events are acceptable.
4. Feeders between Class 1E power systems and systems located in nonsafety class structures should be provided with Class 1E circuit breakers or fuses located in a safety class structure.
5. Class 1E system equipment should be qualified for its intended function under the stipulated conditions.
6. Undue interaction between Class 1E and Non-Class 1E systems should not be possible. When it is necessary to establish a connection between Class 1E and Non-1E, this duration should be as short as possible and should be justified only for specially analyzed cases.
7. In addition of controls and indication in the control room, the plant design should control and indication outside the main control room for circuit breakers that switch Class 1E buses between the preferred power source and the standby power supply, and any other equipment required for safety systems that must function to bring the plant to a safe shutdown condition.
8. Unique identification is essential for Class 1E systems. All related documents should be marked and/or labeled in a distinctive manner.
9. The Class 1E systems should be designed in compliance with the single failure criterion (SFC) [7], to perform all safety functions required for a design basis event in the presence of any single detectable failure within the Class 1E power systems concurrent with all identifiable but nondetectable failures, all failures caused by the single failure, and all failures and spurious system actions that cause or are caused by the design basis event requiring the safety functions. The single failure could occur prior to, or at any time during, the design basis event for which the safety system is required to function. The performance of a probabilistic assessment of the Class 1E power system may be used to demonstrate that certain postulated failures need not be considered in the application of the criterion. Performance of a probabilistic assessment helps to eliminate consideration of events and failures that are not credible; it should not be used in lieu of the single-failure criterion.
10. Connection of non-Class 1E circuits to Class 1E power systems should, in general, be discouraged. However, if connections are strictly necessary, they should be made such that the Class 1E systems will not be degraded.
11. Surveillance and testing should be provided of all automatic and manual transfers of power sources, including simulation of the loss of off-site power in conjunction with a safety injection actuating signal [3].

4.2 Specific Requirements for Class 1E ac Power Systems

The ac power system should include power supplies and distribution systems arranged to provide power to the Class 1E ac loads and controls. Features such as physical separation, electrical isolation, redundancy, and qualified equipment need to be included in the design to aid in preventing a mechanism by which a single design basis event could cause redundant Class 1E equipment within the station to be inoperable.

Particular requirements for the Class 1E ac Power Systems are as follows:

a) The degradation of the Class 1E power systems below an acceptable level should be precluded by providing adequate protective devices. The protective actions of each load group should be independent of the protective actions provided by redundant load groups.
b) Special arrangements need to be incorporated in the design of the standby power supply so that any design basis event will not cause failures in redundant power sources. In addition, the design needs to minimize common-cause failures of a preferred power source and standby power source associated with a single load group (refer to Chapter 6 for detail requirements).
c) The duration of the connection between the preferred power supply and the standby power supply should be minimized (e.g., limited to the time required to perform standby power supply testing) (refer to IEEE Std 741 for information on automatic bus transfers that may be included in the design of these systems).

4.3 Specific Requirements for Class 1E DC Power Systems

The DC power systems include power supplies and distribution systems arranged to provide power to the Class 1E DC loads and for control and switching of the Class 1E Power Systems. As with ac Class 1E Power systems, features such as physical separation, electrical isolation, redundancy, and qualified equipment should be included in the design to aid in preventing a mechanism by which a single design basis event can cause redundant equipment within the station Class 1E power system to be inoperable. (For more information on DC power systems, refer to Chapter 7 "DC Energy Storage and Distribution System.")

All station batteries should be periodically checked in accordance with specifications to provide an indication of battery cells becoming unserviceable before they fail.

4.4   Specific Requirements for Class 1E Instrumentation and Control Systems

The instrumentation and control power systems include power supplies and distribution systems arranged to provide ac and/or DC electric power to the Class 1E Instrumentation and Control loads.

These systems are to be designed to provide a highly reliable source of power to the reactor trip system, engineered safety features, auxiliary supporting features, and other auxiliary features.

The sources and effects of harmonics are to be considered. To accomplish this requirements, special power supplies may be needed that are isolated from the ac and DC power supplies used for the normal instrumentation and control of the unit(s).

Two or more independent DC power supplies should be provided for instrumentation and control. Within each redundant division, the DC source may be a common battery for both Class 1E DC power and instrumentation and control loads.

Two or more independent ac power supplies should be provided for instrumentation and control.

Surveillance to monitor the status of the instrument and control power system ac supply indicators should be provided. This instrumentation should include indication of the following:

- output voltage,
- output current,
- circuit breaker/fuse status, and
- distribution bus frequency.

The execute features should include actuation devices, interconnecting wire and cabling, and actuated equipment that utilize electric power to provide actions when signals are received from the sense and command features. If manual control of any actuated equipment is required, the features necessary to accomplish such manual control should be Class 1E.

Protective devices should be provided for the actuated equipment to limit degradation of the Class 1E actuated equipment. Sufficient indication should be provided to identify the actuation of the protective device. Where application of the protective devices can prevent completion of a safety function, they may be omitted (or bypassed), provided such omission does not degrade the Class 1E power system below an acceptable level [2].

Surveillance and test requirements should be provided in accordance with the requirements of Criterion XIV, "Inspection, Test, and Operating Status," of Appendix B to 10 CFR Part 50 [6].

Criterion XIV establishes that measures be established for indicating the operating status of structures, systems, and components of the nuclear power plant, such as by tagging valves and switches, to prevent inadvertent operation. Section 50.55a, "Codes and Standards," of 10 CFR Part 50, requires in Paragraph (h) that protection systems meet the requirements set forth in the Institute of Electrical and Electronics Engineers "Criteria for Nuclear Power Plant Protection Systems" (IEEE 279). Section 4.13 of IEEE Standard 279, "Criteria for Protection Systems for Nuclear Power Generating Stations," (also designated ANSI N42.7) requires that, if the protective action of some part of the protection system has been bypassed or deliberately rendered inoperative for any purpose, this fact should be continuously indicated in the control room. Applicable Nuclear Regulatory Commission position is provided in Regulatory Guide 1.47.

Protective devices should be provided for the actuated equipment of the execute features to limit degradation of the Class 1E actuated equipment. Sufficient indication should be provided to identify the actuation of the protective device. Where application of the protective devices can prevent completion of a safety function, they may be omitted (or bypassed), provided such omission does not degrade the Class 1E power system below an acceptable level.

4.5 Specific Requirements for Class 1E Containment Electrical Penetrations

Containment electrical penetrations are special devices designed to carry the cables and conductors that must penetrate containment. They are designed to preclude a breach of the containment integrity and need to comply with special electrical, mechanical, and structural requirements. The containment electrical penetrations are discussed in detail in Chapter 5.

Failure of any circuit that penetrates containment should not result in exceeding the current versus time capability of the containment penetration for that circuit during normal operation or during any design basis event requiring containment isolation.

4.6 Specific Requirements for Emergency On-Site ac Power Sources

Emergency on-site ac power sources are located inside the plant in safety-related structures, and they must be ready to supply power when the off-site source fails. These systems are discussed in detail in Chapter 6.

Questions and Problems

4.1   Provide the safety classification that must be assigned to the breakers connecting the preferred power supply to the Class 1E on-site distributions system. Give an explanation for your answer.

4.2   Is the single failure criterion replaceable by a probability analysis that indicates that a single failure will not occur?

4.3   Provide the main information that should be included in a design basis document for a nuclear plant Class 1E ac system.

4.4   Define the applicability of the single failure criterion to the Class 1E system.

4.5   Provide the safety reasoning for discouraging Non-Class 1E connections to the Class 1E systems.

4.6   Provide an acceptable approach for connecting Non-Class 1E equipment to the Class 1E systems.

4.7   Assume a nuclear plant where there is no redundancy for a safety-related pump. Is it possible to postulate plant compliance with the single failure criterion? Justify your answer.

4.8   Assume that a failure mode analysis has identified a lack of independence between the two redundant emergency diesel generators, such that compliance with the single failure criterion would be challenged. The plant requested approval for the plant design on the basis of a probability analysis that determined the likelihood of the loss of independence would be very small. Is the plant approach correct? Justify your answer.

4.9   A plant safety-related service water pump feeder cable failed due to water immersion for extended time. The redundant pump did not fail but its feeder cable was judged to be prone to water immersion as well. What should the plant corrective measure include? Justify your answer.

4.10   An unusual flood event disabled the access roads to a plant, but did not affect the operation of the plant, transmission lines/switchyard, nor any safety-related system. The roads would take 3 months to be repaired. The plant could only be resupplied of emergency diesel generator fuel via access roads (now disabled). Since the plant area was seriously deficient on energy, it was suggested that the plant could be restarted and continue operation to provide energy to the site. Enumerate the possible risks of the suggestion.

References

1 IEEE 279, "Criteria for Protection Systems for Nuclear Power Generating Stations,'" (also designated ANSI N42.7).
2 Regulatory Guide 1.47, "Bypassed and Inoperable Status Indication for Nuclear Power Plant Safety Systems," U.S. Nuclear Regulatory Commission, Washington, DC.
3 CFR (Code of Federal Regulations), Title 10: Energy, Part 100, Criterion XIV of Appendix B, "Inspection, Test, and Operating Status."
4 IEEE 741, "IEEE Standard Criteria for the Protection of Class 1E Power Systems and Equipment in Nuclear Power Generating Stations."
5 IEEE 308, "Standard Criteria for Class 1E Power Systems for Nuclear Poser Generating Stations."
6 CFR (Code of Federal Regulations), Title 10: Energy, Part 100, published by Office of the Federal Register.
7 IEEE 379, "IEEE Standard Application of the Single-Failure Criterion to Nuclear Power Generating Stations."

5

Nuclear Plants Containment Electrical Penetration Assemblies

5.1 Containment Electrical Penetration Assemblies: General (Information on this chapter is based on the requirements of IEEE 317)

Nuclear power plant containment structures are entirely sealed-off constructions containing the nuclear reactor. The containment is designed to preclude leakage of radiation to the environment.

Electrical penetration assemblies (EPAs) perform two key functions:

a) They provide the pass-through for power, control, and instrumentation cables to the thousands of instruments, control panels, electric motors, and many other electric and electronic devices located within the containment.
b) They maintain the pressure boundary integrity of the containment structure.

EPAs are hence critical safety components, and reliability of the device is of utmost importance. This chapter is based on the requirements of IEEE 317 [19] elaborating on the design, construction, qualification, test, and installation, of electric penetration assemblies in nuclear containment structures for stationary nuclear power generating stations.

Canister and modular designs are available for EPAs, with a variety of electrical and fiber optic channels (called feedthroughs). All electrical feedthroughs contain solid copper conductors. Design life is 60 years, normal service temperature is 150°F (65°C)—higher temperatures can be accommodated with an adjustment in the current rating. All EPAs are qualified to plant-specific parameters that include temperature, radiation, pressure, and chemical spray, to comply with the requirements of IEEE 317. All EPAs are ASME Boiler and Vessel Code (NPT Stamp) for manufacture to Section III, Subsection NE, Class MC, and are qualified by test to the current standards of IEEE 317, IEEE 323, IEEE 344 [6]. All parts are manufactured to ANSI/ASME NQA-1 [2] and 10.

Electrical Systems for Nuclear Power Plants, First Edition. Omar S. Mazzoni.
© 2019 by The Institute of Electrical and Electronic Engineers, Inc. Published 2019 by John Wiley & Sons, Inc.

Figure 5.1 Containment electrical penetration for low voltage cables. *Mounting type*: by flange and a *pressure gage* is shown connected to the flange.

A typical EPA is shown in Figure 5.1 (from Conax). Other designs are also available, including epoxy design. To avoid any ingress of moisture into the free space of the EPA along cables, they are sealed longitudinally. The end-flange and the integrated metallic components of the cable glands are made of stainless steel. All sealing and cables must be loss-of-coolant accident (LOCA)- proof.

The cable jointing box covering the end region of each instrumentation and control EPA provides a further barrier able to resist the effects of moisture, high temperature, and increased pressure in an accident. The metal enclosure is made of stainless steel, resistant to chemical attack.

The penetration assembly should be installed, inspected, and tested in accordance with ASME Boiler and Pressure Vessel Code, Division 1, Section III, Subsection NE for Class MC Components [9], and in accordance with applicable parts of IEEE 336 [5]. After installation of an EPA, the penetration module is filled with N_2 gas. The internal space of the penetration is connected to a pressure gauge for continuous leak rate testing. Alarms are provided for low gas pressure.

5.2 Service Classification

The conductors of each circuit of an electric penetration assembly should be assigned one of the following service classifications based on its use:

- medium voltage power,
- low voltage power,
- control voltage, and
- instrumentation.

Medium Voltage Power Penetrations

Penetrations with conductors of power circuits having rated values above 1000 V [16] are classified as *medium voltage power* and should include specification of the following ratings during normal and during the most severe design basis events (DBE) environmental conditions:

1. rated voltage,
2. rated continuous current,
3. rated short time overload current/duration, and
4. rated short circuit thermal capability.

Low Voltage Power Penetrations

Penetrations with conductors of power circuits rated 1000 V and below are classified as *low-voltage power* and should have the following ratings specified under normal conditions and the most severe DBE conditions:

1. rated voltage,
2. rated continuous current,
3. rated short time overload current/duration, and
4. rated short circuit thermal capability.

Control Voltage Penetrations

Penetrations with conductors of control circuits rated 1000 V and below are classified as *control* and should have the following ratings specified under the most severe DBE conditions:

1. rated voltage,
2. rated continuous current,
3. rated short circuit thermal capability,
4. rated short time overload current/duration.

Instrumentation Penetrations

Conductors of instrumentation circuits should have a voltage rating defined by the design service conditions. Included under instrumentation conductors, they are, for example, coaxial, triaxial, resistance temperature detector (RTD), thermocouple, and twisted-shielded circuits.

Optical Fibers Penetrations

Optical fibers have no electrical voltage or current rating. Fibers may be installed in all four (4) types of the electric penetration electrical service classifications, with instrumentation type electric penetrations being the preferred location. Fibers should have the following ratings specified under normal and under the most severe DBE conditions:

- mode type, either single- or multimode,
- index, step, or graded,
- core size, and
- attenuation decibel loss, see [17].

5.3 Electrical Design Requirements (extracted from IEEE 317)

General

Medium voltage power conductors should be free of partial discharge (corona) when energized at the maximum operating voltage.

Instrumentation circuits should be designed to meet the requirements defined by the design service conditions.

The insulation system of each conductor should be capable of withstanding the dielectric tests without exhibiting any signs of deterioration.

Thermocouple conductors and connections should be designed within the error limits for thermocouple circuits in accordance with the latest version of ISA MC 96.1. [1]

Connections and splices of medium and low voltage power and control conductors should be capable of carrying rated continuous current prior to and following rated short circuit current without the temperature of the connections or splices exceeding the rated temperature of the conductor or the connection failing.

Electric penetration assemblies having conductors of more than one voltage rating in the assembly should be designed with a ground barrier separating the conductors of each voltage rating.

The heat generated by hysteresis and eddy current losses should be included in the design requirements for power and control conductors.

Requirements to be met include:

- rated continuous current
- rated voltage
- thermal cycling due to normal operation
- rated short time overload current and duration
- rated short circuit current

- rated short circuit thermal capacity
- ability to withstand design basis events including LOCA, high energy line break (excluding direct steam jet impingement), seismic events, and other designated events.

Electrical Integrity

The assemblies should pass a test for electromagnetic compatibility (EMC) for emissions or susceptibility.

Furthermore, the conductors, connections, and electrical insulation systems should be designed to withstand without failure or loss of function the relevant design service conditions.

Rated Voltage

Penetrations are rated as follows:

- 8000 V, with conductors rated 5001–8000 V
- 5000 V, with conductors rated 1001–5000 V
- 1000 V, with conductors rated 601–1000 V
- 600 V, with conductors rated 301–600 V
- 300 V, with conductors rated up to 300 V

Rated Continuous Current

The rated continuous current in amperes is that which a conductor can carry continuously without the stabilized temperatures of the conductor and the penetration nozzle–concrete interface (if applicable) exceeding their design limits (see NFPA-70 [8] and IEEE 317 [19]), whereas all other conductors in the assembly also carry their rated continuous current, under the maximum normal environment temperature imposed by the design.

Rated Short Time Overload Current/Duration

The rated short time overload current and duration should be the overload current in amperes that each conductor of a circuit can carry for a specified duration, following continuous operation at rated continuous current, without the temperature of the conductors exceeding their short time overload design temperature limit with all other conductors in the assembly carrying their rated continuous current under the maximum normal environment temperature of the design service conditions (see NFPA-70 [8] and IEEE 317 [19]). The rated short time overload current should not be less than seven times the rated continuous current of the conductor, and the duration should be not less than 10 s.

Rated Short Circuit Current

The rated short circuit current should be the current in amperes that each conductor of a circuit can carry while maintaining electrical integrity with all other conductors in the assembly carrying their rated continuous current under the maximum normal environment temperature of the design service conditions (see NFPA-70 [8], ICEA P-32-382 [11], and IEEE 317 [19]).

The rated short circuit current for alternating-current circuits should be expressed in rms symmetrical amperes.

The rated short circuit current for DC circuits should be based on a current evidencing a constant DC value.

Rated Short Circuit Thermal Capacity

The rated short circuit thermal capacity should be the product of short circuit current in amperes squared (see NFPA-70 [8] and IEEE 317 [19]), and its duration in seconds that each conductor of a circuit can carry following continuous operation at rated continuous current while maintaining electrical integrity with all other conductors in the assembly carrying their rated continuous current under the maximum normal environment temperature of the design service conditions.

The rated short circuit thermal capacity should be expressed in amperes squared seconds ($I^2 t$) subject to the following limits:

- The available short circuit current should not exceed the rated short circuit current and should be expressed in rms symmetrical amperes for ac circuits and DC amperes for DC circuits.
- The design should include a maximum duration of the short circuit current of 2 s.

Rated Continuous Current During the Most Severe DBE Environmental Conditions

For rating of conductors, refer to NFPA -70 [8] and IEEE 317 [19]. The conductors in the electric penetration assembly should have a rated continuous current during the most severe DBE environmental conditions not less than the rated continuous current in for which the conductors should maintain electrical integrity and containment integrity.

Rated Short Time Overload Current and Duration during the Most Severe DBE Environmental Conditions

For rating of conductors, refer to NFPA -70 [8] and IEEE 317 [19]. The conductors in the electric penetration assembly should have a rated short time overload current and duration during the most severe DBE environmental conditions, meeting the following conditions:

- not less than the rated short time overload current and duration for which the conductors maintain electrical integrity and containment integrity,
- with the remaining conductors in the assembly carrying rated continuous current, and
- the conductors should be capable of meeting the above requirements at the maximum conductor temperature attained during the most severe DBE environmental conditions.

Rated Short Circuit Current during the Most Severe DBE Environmental Conditions

The conductors in the electric penetration assembly should have a rated short circuit current during the most severe DBE environmental conditions not less than the rated short circuit current for which the conductors should maintain containment integrity with the remaining conductors in the assembly carrying rated continuous current and without affecting the mechanical and electrical integrity of the remaining conductors. The conductors should be capable of meeting the above requirements at the maximum conductor temperature attained during the most severe DBE service conditions.

Rated Short Circuit Thermal Capacity ($I^2 t$) during the Most Severe DBE Environmental Conditions

The conductors in the electric penetration assembly should have a rated short circuit thermal capacity ($I^2 t$) during the most severe DBE environmental conditions not less than the rated short circuit thermal capacity ($I^2 t$), following continuous operation at rated continuous current while maintaining containment integrity of the assembly and without affecting the electrical integrity and the mechanical integrity of the conductors not subjected to the short circuit current. The conductors should be capable of meeting the above requirements at the maximum conductor temperature attained during the most severe DBE environmental conditions. The short circuit thermal capacity ($I^2 t$) should be expressed in ampere squared-seconds subject to the following limits:

- The short circuit current should not exceed its rated value.
- The duration should not exceed 2 s.
- The short circuit current should be expressed in rms symmetrical amperes for ac circuits and DC amperes for DC circuits.

5.4 Mechanical Design Requirements (Extracted from IEEE 317)

Pressure Boundary

The mechanical design, materials, fabrication, examinations, and testing of the pressure boundary of the electric penetration assembly should be in accordance with the requirements of the ASME Boiler and Pressure Vessel Code, Division 1, Section III, Subsection NE for Class MC Components.

Stress calculations should include stresses from all applied loads including electromagnetic forces produced by rated short circuit currents.

Design Pressure and Temperature

The design pressure and temperature should be determined and specified in accordance with the rules of the ASME Boiler and Pressure Vessel Code, Division 1, Section III, Subsection NE for MC containment vessels [1]. NOTE: Under severe accident conditions (SAC), the containment may be subjected to higher pressure and temperature. Consideration may be given to qualify the electric penetration to a pressure rating comparable to the containment rating to prevent leakage paths for the severe accident environment and preserve containment integrity.

Minimum Design Temperature

The minimum design temperature of the electric penetration assembly should be −28°C (−20°F).

Design Gas Leak Rate

The electric penetration assembly, exclusive of the aperture seal(s), should be designed to have a total gas leak rate not greater than 1×10^{-3} standard cm^3/s of dry nitrogen at 20 ± 15°C (68 ± 27°F) at the design pressure.

The aperture seal(s) of the electric penetration assembly should be designed to have a total gas leak rate not greater than 1×10^{-3} standard cm^3/s of dry nitrogen at 20 ± 15°C (68 ± 27°F) at the design pressure.

The electric penetration assembly, including the aperture seal(s), should be designed to have a total gas leak rate not greater than 1×10^{-2} standard cm^3/s of dry nitrogen at the design pressure and temperature.

Gas Leak Rate Testing and Monitoring Provisions

Electric penetration assemblies having double electric conductor seals or optical fiber seals or double aperture seals, or both, should be designed for and should include provisions for gas leak rate testing and monitoring of the double seals after installation.

Valves, fittings, and pressure piping and gauge(s) provided for this purpose should have a pressure rating not less than 110% design pressure.

Electric penetration assemblies should include provisions for gas leak rate monitoring during storage.

Mechanical Integrity

Under design service conditions, the mechanical support systems, conductors, fibers, terminations and conductor support systems, pressure barrier(s), and conductor, fiber, and aperture seals should be designed to withstand the following conditions without failure or loss of function:

- rated continuous current,
- thermal cycling due to normal operation,
- rated short time overload current and duration
- rated short circuit current,
- rated short circuit thermal capacity (I^2t), and
- design basis events including LOCA, high energy line break (excluding direct steam jet impingement), seismic events, and other designated events.

Containment Integrity

The electric penetration assembly including aperture seal(s) should be designed to have a total gas leak rate not greater than 1×10^{-2} standard cm^3/s using dry nitrogen at design pressure and ambient temperature after installation and **after** any design basis events. Storage should be in accordance with Level B packaging should be in accordance with Level C or better.

Electric penetration assemblies having double electrical conductor seals or optical fiber seals or double aperture seals, or both, that are completely assembled and that require protection from exposure to the atmosphere, should have the internal parts of the sealed system maintained at a positive pressure and should include monitoring provisions.

5.5 Fire Resistance Requirements (Extracted from IEEE 317)

Fire resistance requirements should be met, as follows:

- Insulated cables and splices should be qualified in accordance with IEEE 383 [7].
- All other organic, nonmetallic material, except those materials that by their physical configuration will not contribute to a fire, should meet at least one of the following requirements as applicable:
  1. Classified as *nonburning* or *self-extinguishing* in accordance with ASTM D 635-10 [10], or,
  2. Classified as having an *oxygen index* not less than 26 in accordance with ANSI/ASTM D2863-10 [3].

5.6 Qualified Life

The electric penetration assembly should be designed to have a qualified life not less than the installed life. If the qualified life of an electric penetration subcomponent(s) is less than the installed life, as determined by the qualification, replacement of the subcomponent(s) should be required, prior to the end of its qualified life, or the qualified life extended in accordance with IEEE 323 [14].

5.7 Qualification Tests

Qualification should be established by tests that demonstrate that the electric penetration assembly will perform its intended function when subjected to conditions that simulate installed life under the design service conditions.

Qualification [13], [14], [15] should include

- *design tests* that demonstrate the adequacy of the design to meet established requirements and
- *qualified life tests* that address aging and establish the qualified life of the assembly.

Design tests may be performed in any sequence on different test specimens. Qualified life tests should be performed on preconditioned test specimens.

Documented analyses, with justification of methods, theories, and assumptions or additional testing of components or assemblies, or both, may be used with existing qualification tests to qualify the design to service conditions that differ from the design service conditions under that qualification was demonstrated.

Each conductor or fiber size, rating, and configuration need *not* be subjected to qualification tests; instead, representative test specimens may be used.

5.8 Design Tests (Extracted from IEEE 317)

The following design tests should be performed for each generic design:

Gas Leak Rate Test

The gas leak rate of the test specimen, including nonwelded aperture seals, should not exceed a total leak rate equivalent to 1×10^{-2} standard cm^3/s of dry nitrogen when tested at design pressure and temperature.

Pneumatic Pressure Test

Where the electric penetration assembly is designed as a pressure vessel in accordance with the ASME Boiler and Pressure Vessel Code, it should be pneumatically pressure tested in accordance with Division 1, Section III, Subsection NE, Article NE-6000 of the code [9].

Dielectric Strength Tests

The following dielectric strength tests should be conducted at test facility room ambient conditions of temperature, pressure, and relative humidity:

Power Frequency Voltage Test. Each medium voltage power, low voltage power, and control conductor should be given a 60-Hz sinusoidal voltage test for not less than 1 min applied between each conductor and ground and between each conductor and adjacent conductors not separated by a ground barrier. The test voltage should be based on the voltage rating of the conductor in accordance with the following table (ref IEEE317) [19]:

Insulation rated voltage (V)	RMS test voltage (V)
300	1600
600	2200
1000	3000
5000	19,000
8000	36,000
15,000	36,000

Impulse Voltage Tests

Each medium voltage power conductor should be given a full-wave 1.2×50 µs impulse voltage test series, with a crest voltage not less than the following (ref IEEE 317)

Insulation-rated voltage (V)	Impulse crest voltage (kV)
5000	60
8000	95
15,000	95

The impulse voltage tests and acceptance criteria should be in accordance with IEEE/C37.23, 5.2.1.2 [4].

Instrumentation conductors should be tested to demonstrate that they meet the requirements of the design service conditions.

Insulation Resistance Test

The insulation resistance test should be performed at test facility room ambient conditions of temperature, pressure, and relative humidity.

Medium voltage power conductors should be tested at 500 V dc (minimum) and should have a minimum resistance of 1000 $M\Omega$ between the conductor and ground and between the conductor and adjacent conductors not separated by a ground barrier

Low voltage power and control conductors should have a minimum resistance of 100 $M\Omega$.

Instrumentation conductors should be tested to demonstrate that they meet the requirements of the design service conditions.

Partial Discharge (Corona) Test

Medium-voltage power conductors should be tested for partial discharge (corona) between phase conductors and should have a partial discharge extinction voltage not less than rated voltage, at ambient temperature and humidity. The test apparatus, calibration, and test procedure should be in accordance with ICEA S-93-639/ NEMA WC 74-2006, Clause 9.8.2. [16]

Rated Continuous Current Test

The test specimen carrying its rated continuous current under the maximum normal environment temperature of the design service conditions.

The maximum stabilized temperatures should not exceed the design limits.

Rated Short Time Overload Current Test

Each size conductor of the test specimen should be verified by testing under the maximum normal environment temperature of the design service conditions. The maximum temperatures of the overloaded conductors attained during the test should be recorded and should not exceed the design limits for each conductor.

Rated Short Circuit Thermal Capacity (I^2t) Test

The short circuit thermal capacity (I^2t) of each size conductor of the test specimen should be verified by testing refer to IEEE 317, Table A.3.

Seismic Test

The test specimen should be seismically qualified in accordance with IEEE Std 344 [6] for the input motion spectra in the design service conditions, plus margin. Testing should be performed under conditions that simulate the installed assembly including consideration of terminal boxes, external cables, and raceways. During the test, all conductors in the test specimen should maintain uninterrupted current continuity and should withstand rated voltage plus 10% margin.

Installation Welding Test

Where the method of penetration attachment is by welding, a test should be conducted to demonstrate that a representative electric penetration assembly can be welded into the containment vessel without damage by following the manufacturer's recommended procedures.

After the welding test, the test specimen should be examined for signs of physical damage and pass the leak rate test as specified in 8.3 of iEEE 317 and the electric tests.

Electro-Magnetic Compatibility Test

An EMC emissions test may be performed on representative conductors of medium voltage power and low voltage power type electric penetrations. The electromagnetic interference (EMI)/radio frequency interference (RFI) susceptibility test may be performed on representative conductors of instrumentation type electric penetrations. Test parameters may be established from IEEE-603-2009 Annex B [18].

Qualified-Life Tests

Qualified life tests should be performed for each *thermal operating cycle simulation*. The test specimen should be subjected to not less than 120 cycles of temperature changes in the specimen of not less than 55°C (100°F) for each cycle. One cycle equates a temperature increase of 100°F and a decrease of 100°F.

- *Thermal Age Conditioning.* The test specimen should be thermally age conditioned to simulate operation at design normal service temperature for the installed life. Accelerated thermal aging time and temperature derived can be derived from Arrhenius data (using procedures in accordance with IEEE 98 [12] and IEEE 101 [13].

- *Radiation Exposure Simulation.* The test specimen should be exposed to radiation simulating the design normal service environment radiation for the installed life, unless it can be demonstrated that radiation does not degrade the test specimen (see IEEE 323) [14].
- *Short Circuit Current and Short Circuit Thermal Capacity Tests.* The short circuit current and the short circuit thermal capacity of each size conductor should be verified by testing.
- *Seismic Test.* The test specimen should be seismically qualified.
- *Simulation Tests of the Most Severe DBE Environmental Conditions.* One test specimen (or specimens if more than one is required to represent a generic design) should be exposed to design loss-of-coolant-accident design service conditions, and another test specimen (or specimens if more than one is required to represent a generic design) should be exposed to design high-energy-line-break design service conditions, excluding the direct impingement of steam jets on the test specimen. Additional test specimens should be exposed to the service conditions of other DBEs if they produce more severe environmental service conditions than loss-of-coolant accident and high energy line break. The loss-of-coolant accident and high energy line break tests that simulate individual DBEs may be combined into one test provided the most severe design service conditions, including pressures, temperatures, humidity, radiation (if not included in the preconditioning), and chemical or demineralized water sprays, and submergence (if required), are used in the combined test.

Accelerated thermal testing may be used to simulate the temperature-time profile following the major temperature transient(s) of the most severe DBE environmental conditions.

The test should be conducted with a sufficient number of conductors carrying continuous test current to produce the same total conductor I^2R heating effects as is produced by rated continuous current in all the conductors in the assembly. Not less than 20% of the conductors should be continuously energized at rated voltage during the test.

At least one of each type of instrumentation circuit should have the insulation resistance or leakage current recorded continuously during the test, except for interruptions as may result from switching or calibration for data acquisition. The remainder of the instrumentation circuits should be energized at rated voltage and their insulation resistance or leakage current measured periodically. The insulation resistance or leakage current should be within design service condition limits.

After the test, the specimen should be given the leak rate test. Where the test specimen does not include the aperture seal, the test specimen should pass the leak rate test as specified in 8.3 of IEEE 317, the leak rate should not exceed 1×10^{-3} standard cm^3/s of dry nitrogen.

Determining Qualified Life

If individual components of a design have a qualified life less than the required installed life of the design, periodic replacement of these components may be used to extend the qualified life of the design to the required installed life.

5.9 Production Tests

The following production tests should be performed on each penetration assembly prior to shipment:

- gas leak rate test,
- pneumatic pressure test,
- dielectric strength test,
- insulation resistance test,
- conductor continuity and identification tests,
- partial discharge (corona) test, and
- optical fibers should be tested in accordance with the manufacturer's recommendations.

5.10 Monitoring and Testability

In accordance with IEEE 603, periodic testing of the penetrations is required to ascertain that they remain functional. Leak rate monitoring should be performed of the penetration seals, by conducting pressure testing at the seals.

The rate of gas leakage must remain below the level defined by relevant standards, to prevent the escaping of radioactive by-products into the environment.

Questions and Problems

5.1 What electrical protective relay features are provided to prevent an electrical penetration assembly failure?

5.2 What are the immediate consequences of an electrical penetration assembly failure?

5.3 Provide a brief report on the postulated failure of electrical penetration assemblies at Fort Calhoun Nuclear Plant (consult the NRC website and other relevant sources). Comment on any specification, testing, or qualification shortcomings may have been the reasons for the postulated failures.

5.4 Evaluate a failure of an electrical penetration assembly that could potentially result in a breach of the nuclear reactor containment, and relate to the requirements of the single failure criterion. Provide a rational for your response.

5.5 What are the preventive measures instituted for assuring continuing reliability of electrical penetration assemblies?

References

1 ISA MC 96.1, "Temperature Measurement Thermocouples."
2 ASME NQA-1, "Quality Assurance Requirements for Nuclear Facility Applications."
3 ANSI/ASTM D2863-10, "Standard Method for Measuring the Minimum Oxygen Concentration to Support Candle-Like Combustion of Plastics (Oxygen Index)."
4 IEEE C37.23, "IEEE Standard for Metal-Enclosed Bus."
5 IEEE Std 336, "IEEE Guide for Installation, Inspection, and Testing Requirements for Class 1E Power, Instrumentation and Control Equipment at Nuclear Facilities."
6 IEEE Std 344, "IEEE Recommended Practice for Seismic Qualification of Class lE Equipment for Nuclear Power Generating Stations."
7 IEEE Std 383, "IEEE Standard for Qualifying Class lE Electric Cables and Field Splices for Nuclear Power Generating Stations."
8 NFPA-70, "National Electrical Code."
9 ASME Boiler and Pressure Vessel Code.
10 ASTM D635-10, "Standard Test Method for Rate of Burning and/or Extent and Time of Burning of Self-Supporting Plastics in a Horizontal Position."
11 ANSI/ICEA P-32-382, "Short-Circuit Characteristics of Insulated Cables."
12 IEEE Std 98, "IEEE Standard for the Preparation of Test Procedures for the Thermal Evaluation of Solid Electrical Insulating Materials."
13 IEEE Std 101, "IEEE Guide for the Statistical Analysis of Thermal Life Test Data."
14 IEEE Std 323, "IEEE Standard for Qualifying Class lE Equipment for Nuclear Power Generating Stations."
15 IEEE Std 627, "IEEE Standard for Qualification of Equipment Used in Nuclear Facilities."
16 ICEA S-93-639/ NEMA WC 74, "5–46 kV Shielded Power Cable for Use in the Transmission and Distribution of Electric Energy, Clause 9.8."

17 ANSI EIA 455-171, "Attenuation by Substitution Measurement for Short-Length Multimode Graded-Index and Single-Mode Optical Fiber Assemblies."

18 IEEE Std 603, "IEEE Standard Criteria for Safety Systems for Nuclear Power Generating Stations."

19 IEEE 317, "IEEE Standard for Electric Penetration Assemblies in Containment Structures for Nuclear Power Generating Stations."

6

On-Site Emergency Alternating Current Source

6.1   General Requirements of the Emergency Alternating Current Source

While gas turbine-driven and hydraulic-driven generators may be considered as the on-site emergency source, most plants utilize diesel generators as the chosen on-site emergency power source. One noted exception is the Oconee Nuclear Plant, operated by Duke Power, which utilizes a hydroelectric plant as an emergency power source.

Diesel generators (Figure 6.1) are generally selected as emergency standby units, particularly because they provide the best starting time and load energization response as required to mitigate an accident (Applicable standards are: IEEE 308 [16], IEEE 603 [21], ANS59.51 [12], ANSI C50.10 [13], ANSI/ASME Pressure Vessels [14], and ANSI/NFPA 37 [15]).

Applicable regulations for the emergency ac source are:

General Design Criterion 17, "Electric Power Systems," of Appendix A, "General Design Criteria for Nuclear Power Plants," to Title 10, Part 50, of the Code of Federal Regulations (10 CFR Part 50), "Domestic Licensing of Production and Utilization Facilities" [1], requires that *on-site electric power systems* have sufficient independence [20], capacity, capability, redundancy, and testability to ensure that

1. specified acceptable nuclear fuel design limits and design conditions of the reactor coolant pressure boundary *are not exceeded* as a result of anticipated operational occurrences and
2. the core is cooled and containment integrity and other vital functions are maintained in the event of postulated accidents, assuming a single failure.

General Design Criterion 18, "Inspection and Testing of Electric Power Systems," of Appendix A to 10 CFR Part 50 requires that electric power systems important to safety be designed to permit appropriate periodic inspection

Electrical Systems for Nuclear Power Plants, First Edition. Omar S. Mazzoni.
© 2019 by The Institute of Electrical and Electronic Engineers, Inc. Published 2019 by John Wiley & Sons, Inc.

Figure 6.1 Typical diesel generator set.

and testing to assess the continuity of the systems and the condition of their components.

Criterion III, "Design Control," and Criterion XI, "Test Control," of Appendix B, "Quality Assurance Criteria for Nuclear Power Plants and Fuel Reprocessing Plants," to 10 CFR Part 50 requires that

- measures be provided for verifying or checking the adequacy of design through design reviews, the use of alternative or simplified calculational methods, or the performance of a suitable testing program and
- a test program be established to ensure that systems and components perform satisfactorily and that the test program includes operational tests during nuclear power plant operation.

"Loss of All Alternating Current Power" (10 CFR 50.63) requires that each lightwater-cooled nuclear power plant must be able to withstand and recover from a station blackout (i.e., loss of off-site and on-site emergency alternating current (ac) power systems] for a specified duration. *The reliability of on-site ac power sources is one of the main factors contributing to the risk of core melt as a result of a station blackout.*

IEEE 387 provides the general requirements for the design, installation, and testing of emergency diesel generator (EDG) for nuclear power plants.

6.2 General Requirements of Diesel Generators Used as Emergency Alternating Current Source (Information in this chapter is based on the requirements of IEEE 387)

The regulatory position for EDGs is given in *Regulatory guide 1.9,* "Application and Testing of Safety-Related Diesel Generators in Nuclear Power Plants." This guide includes information on the requirements of 10 CFR Part 50 and 10 CFR Part 21, "Reporting of Defects and Noncompliance" [10].

The EDGs must be selected with sufficient capacity, be qualified, and have the necessary reliability and availability for design-basis events.

Specifically, if a loss of off-site power (LOOP) and a design-basis event occur during the same time period, an EDG selected for use in an on-site electric power system should have the capability to

1. start and accelerate a number of large motor loads in rapid succession, while maintaining voltage and frequency within acceptable limits,
2. provide power promptly to engineered safety features, and
3. supply power continuously to the equipment needed to maintain the plant in a safe condition if an extended LOOP occurs (e.g., 30-day period should be considered with refueling every 7 days).

The Institute of Electrical and Electronics Engineers (IEEE) Standard 387, "IEEE Standard Criteria for Diesel Generator Units Applied as Standby Power Supplies for Nuclear Power Generating Stations," delineates principal design criteria and qualification and testing guidelines to ensure that the selected diesel generators meet performance requirements.

EDG General Requirements: Load-Carrying Capability

Knowledge of the characteristics of each load is essential to establish the bases for selection of an EDG that is able to accept large loads in rapid succession. The majority of these emergency loads are large induction motors.

The load profile needs to be developed to show the magnitude and duration of loads applied in the prescribed time sequence, including the transient and steady-state characteristics of the individual loads for the most severe conditions, including both the automatically and manually sequenced loads. The diesel generator unit is typically specified in the design stage, before the actual characteristics of these loads are known. Most of these loads are applied to the diesel generator unit in a combination of block loads, sequenced to best suit the design objectives of the power plant. The majority of these loads are induction motors.

Load Modeling

Knowledge of the characteristics of each load within the block load is essential to establish the rating of the diesel generator unit, a critical application for accelerating large loads in rapid succession.

The following data should be established to properly size the unit:

a) available time to attain rated conditions preceding initial load acceptance,
b) identification of each load block and its application with respect to time sequence

c) type of loads in each load block, for example, transformer, induction motor or resistive load, pump, fan, etc., and

d) load characteristics.

In determining a load profile, it is important to account for the voltage dependence of the loads. The electrical loads are characterized as *constant power* (usually with symbol S), as Constant Impedance (usually with symbol Z), and *constant current* (usually with symbol I).

Typical loads connected to an EDG are transformer loads, inverter loads, motor loads, and resistive loads.

1. *Transformer loads*: These are constant impedance loads and therefore will vary as the square of the voltage at the load terminals. Relevant data include
   i) rated kilovolt amperes, percent impedance (minimum specified), or inrush kilovolt amperes at a rated voltage,
   ii) connected load (power factor and characteristics), and
   iii) efficiency.

2. *Inverter Loads*: These loads are generally constant current loads, thus the change in power is proportional to the voltage change.

3. *Resistive loads*: kilowatts at a rated voltage, these are constant impedance loads, which therefore will vary as the square of the voltage change at the load terminals.

4. *Motor loads*: These are constant kilovolt ampere loads or constant power loads, that is they will represent a fixed kilovolt ampere value as the terminal voltage varies.

Relevant information for motor loads includes (see IEEE 334 [5], and IEEE 382 [7])

i) rated nameplate voltage, frequency, and speed;

ii) rated nameplate horsepower and horsepower at maximum operating conditions; for these conditions include current, power factor, and efficiency;

iii) locked rotor characteristics at rated voltage, including starting power factor and locked rotor current (as percent of full load rating); and

iv) acceleration conditions at rated and minimum required starting voltage, for example, 75%, including starting profile (amperes vs. time).

Note: Depressed voltage below a rated value at the motor terminals, resulting from transformer and distribution losses, will affect motor characteristics and should be identified.

Typical large motor characteristic values are

- horsepower, voltage, and frequency (nameplate),
- locked rotor current (inrush), 6.0–6.5 times rated (approximately 5.5–6.0 kVA/hp),
- locked rotor power factor, 20%,

- breakdown kilowatts, 2.2 HP (typical),
- running efficiency, 92%,
- running power factor, 85%, and
- accelerating time, 2.5–4.0 s (loaded) and 1.5 s (unloaded).

A typical load kW/kVA profile for starting the indicated block loads is usually required. Most load profiles include transformers to be energized. Care should be exercised to ensure that the transformer inrush will not result in an instantaneous voltage dip in excess of acceptable value, so that there will be no degradation of the performance of the design load. Various computer programs are available for analyzing performance of the diesel generator unit under simulated operating and load conditions. Provisions for manually added loads and future load growth should be considered in developing the load profile.

At full voltage, an induction motor draws a starting current of five to eight times its rated full-load current. These sudden large increases in current drawn from the diesel generator as a result of the startup of induction motors can result in substantial voltage reductions. This resultant lower voltage could prevent a motor from starting (i.e., accelerating its load to rated speed in the required time) or could cause a running motor to coast down or stall. Other voltage-sensitive loads might also be lost because of low voltage, or if their associated contactors drop out.

Impact of Frequency Variation
Loading on the EDG is affected by the generator frequency when the generator is operated in the isochronous mode, as the frequency will vary in relationship to the engine governor control accuracy band. The variation in EDG loading can be calculated by use of the affinity laws. The affinity laws are used in hydraulics and heating, ventilation, and air conditioning (HVAC) to express the relationship between variables involved in pump or fan performance (such as head, volumetric flow rate, shaft speed) and power. They apply to pumps, fans, and hydraulic turbines. In these rotary implements, the affinity laws apply both to centrifugal and axial flows.

The affinity laws are useful as they allow prediction of the head discharge characteristic of a pump or fan from a known characteristic measured at a different speed or impeller diameter. The only requirement is that the two pumps or fans are dynamically similar, that is the ratios of the fluid forces are the same.

With impeller diameter (D) held constant:

Law 1a. Flow is proportional to shaft speed:

$$Q_1/Q_2 = N_1/N_2 \tag{6.1}$$

where Q stands for flow and N for speed.

Law 1b. Pressure or head is proportional to the square of shaft speed:

$$H_1/H_2 = (N_1/N_2)^2 \tag{6.2}$$

where H stands for pressure and N for speed.

Law 1c. Power is proportional to the cube of shaft speed:

$$P_1/P_2 = (N_1/N_2)^3 \tag{6.3}$$

Pump speed. An ac motor's synchronous speed, n_s, is the rotation rate of the stator's magnetic field, which is expressed in revolutions per minute as

$$n_s = 120 \times f/p \text{ in rpm} \tag{6.4}$$

where f is the frequency and p is the number of poles.

Impact of Ohmic Losses in the Distribution System
The Ohmic losses in the distribution system need to be evaluated for all cables and transformers fed from the EDG. These loses may amount to 0.5% of the EDG rating, which may become an important addition in cases of low loading margin

Impact of Voltage Variation
Voltage variation, as determined by the accuracy of the voltage regulator, will affect the constant current and constant impedance loads.

The uncertainties inherent in safety load estimates at an early stage of design may be of relative high magnitude, and, therefore, it is prudent to provide a reasonable margin (of at least 5%) in selecting the load capabilities of the EDGs.

As a major factor in reducing the implications of a station black out, the reliability of EDGs is very important among of the main factors affecting the risk of reactor core damage. Nuclear Regulatory Commission (NRC) Regulatory, "Station Blackout" [19] calls for the use of the reliability of the diesel generator as one of the factors in determining the length of time a plant should be able to cope with a station blackout. If all other factors (i.e., redundancy of EDGs, frequency of LOOP, and probable time needed to restore off-site power) remain constant, a higher reliability of the diesel generators will result in a lower probability of a station blackout, with a corresponding decrease in coping duration.

The design of the EDGs should also incorporate high operational reliability, and this high reliability should be maintained throughout their lifetime by implementing a reliability program designed to monitor, improve, and maintain reliability. Increased operational reliability can be achieved through appropriate testing and maintenance, as well as an effective root cause analysis of all EDG failures.

The preoperational and periodic testing data should be used to formulate and back up any necessary corrective actions, so high in-service reliability may be maintained.

EDG General Requirements: Speed Control

In isochronous speed control mode, the energy being admitted to the prime mover is regulated very tightly in response to changes in load which would tend to cause changes in frequency (speed). Any increase in load would tend to cause the frequency to decrease, but energy is quickly admitted to the prime mover to maintain the frequency at the set point. Any decrease in load would tend to cause the frequency to increase, but energy is quickly reduced to the prime mover to maintain the frequency at the set point.

In droop speed control mode, the governor of the prime mover is not attempting to control the frequency (speed) of the ac generator. The term "share the load" causes much confusion, but just refers to the ability of the EDG to smoothly control the production of torque when connected in parallel with the grid.

Droop speed control, in fact, refers to the fact that the energy being admitted to the EDG is being controlled in response to the difference between a speed (frequency) set point and the actual speed (frequency) of the prime mover. To increase the power output of the generator, the operator increases the speed set point of the prime mover, but since the speed cannot change (it is fixed by the frequency of the grid to which the generator is connected) the error, or difference, is used to increase the energy being admitted to the prime mover. So, the actual speed is being "allowed" to "droop" below its set point.

On a small electrical grid, one machine is usually operated in isochronous speed control mode, and any other (usually smaller) generators that are connected to the grid are operated in droop speed control mode. If two prime movers operating in isochronous speed control mode are connected to the same electrical grid, they will usually "fight" to control the frequency, and wild oscillations of the grid frequency usually result. Only one machine can have its governor operating in isochronous speed control mode for stable grid frequency control when multiple units are being operated in parallel. (There are isochronous load sharing schemes in use in various places around the world, but they are not very common.)

On very large electrical grids—commonly referred to as "infinite" electrical grids—all the prime movers are being operated in droop speed control mode, there cannot be any single machine operating in Isochronous speed control mode capable of controlling the grid frequency. But there are so many of them and the electrical grid is so large that no single unit can cause the grid frequency to increase or decrease by more than a few hundredths of a percent as it is loaded or unloaded.

Very large electrical grids require system operators to quickly respond to changes in load to control grid frequency properly since there is no isochronous machine doing so. Usually, when things are operating normally, changes in load can be anticipated and additional generation can be added or subtracted to maintain tight frequency control.

EDG General Requirements: Protection Considerations

An improperly adjusted control system or governor failure may result in the EDG to develop overspeed, which can result tripping of the unit from the overspeed protection, which is usually set at 115% of nominal speed. Likewise, to prevent substantial damage to the generator, the generator differential current trip must operate immediately upon occurrence of an internal fault. Other protective trips can also safeguard the EDGs from possible damage. However, these trips could interfere with successful functioning of the diesel generators when they are most needed (i.e., during design-basis events).

In addition, experience has shown that on numerous occasions, these protective trips have needlessly shut down EDGs because of spurious operation of a trip circuit. Consequently, it is important to take measures to ensure that spurious actuation of these other protective trips does not prevent the EDGs from performing their safety function during the emergency mode of operation.

Owing to the importance to safety of EDGs, many protective features are bypassed under mitigation of a design basis events (DBE). Normally, all electric protection except the generator differential are bypassed under a DBE. Under test and any other operating conditions, all generator protection is not bypassed. The reasoning for the bypass function is to allow for maximum operation under all conditions. In addition, a concern exists for possible spurious operation of protective features. The differential operation is normally very reliable for the detection of internal faults, even though it may not be sufficiently sensitive to detect ground faults closed to the neutral point.

Therefore, to prevent the EDG from providing power under a design bases event, all of the protective systems with the exception of the differential and overspeed protection should be bypassed. For these two trip signals, output signal should be the result of a sufficient number of independent initiating signals in parallel in coincident trip logic. Alternately to bypassing, the trip signal should be the result of a sufficient number of independent initiating signals in parallel (coincident trips). The design of the coincident trip logic circuitry should provide alarm for each individual sensor initiation. During a unit operation in a nonaccident condition, all protective devices should remain in effect see IEEE 741 [23].

Care should be exercised in setting the overcurrent protection of motors to preclude undue tripping of motors due to sustained starting inrush current.

Recovery from the transient caused by starting large motors, or from the loss of a large load, could cause diesel engine to overspeed which, if excessive, might result in a trip of the engine (i.e., loss of the safety-related power source). These same consequences can also result from the cumulative effect of a sequence of more moderate transients if the system is not permitted to recover sufficiently between successive steps in a loading sequence.

The motor torque developed by a typical induction motor is approximately proportional to the square of the voltage, Thus a voltage reduction to 80%, for example, translates into a torque reduction to 64%. To be acceptable, the reduced torque should still be sufficient to accelerate the motor to full speed. Industry practice is to accept a voltage reduction of 10–15% when starting large motors from large-capacity power systems, and a maximum voltage reduction of 25–30% when starting these motors from limited-capacity power sources such as diesel generators. Voltage reduction during load sequencing should also be evaluated in light of the plant-specific equipment to prevent load interruption, for example, contactor coil capability to hold in at lower than nominal voltage. In general, large induction motors, which are not provided with flywheels, can achieve a rated speed in less than 5 s when started from standstill and powered from adequately sized EDGs that are capable of restoring the bus voltage to 90% of nominal in about 1–2 s(s).

6.3   Specific Design Requirements for Emergency Diesel Generators

EDG-specific design requirements apply to the following main component parts:

1. diesel engine,
2. generator,
3. local control, protection, and surveillance systems associated with the EDG, and
4. the ac and DC power supply and distribution systems associated with the EDG.

The diesel engine includes

- flywheel and coupling,
- combustion air system,
- governor system,
- starting system,
- fuel oil system within the EDG room,
- lubricating oil system,
- cooling system, and
- exhaust system.

The generator includes the following:

- main generator leads and
- excitation and voltage regulation.

Each diesel generator should be capable of starting and accelerating to rated speed in the required sequence, all the needed engineered safety features and emergency shutdown loads. The diesel generator unit should have specific capabilities to meet the design, application, and qualification requirements of the IEEE standards and the NRC Regulatory Guides.

The EDG unit should be capable of operating during and after any design basis event without support from the preferred power supply, over the extent of the plant operating licensing period:

- for 4000 starts,
- for a minimum of 6000 h of operation,
- at specified ambient temperature and humidity excursions, and specified barometric pressure (including tornado) and seismic criteria,
- under radiation of 1×104 rad of gamma integrated dose,
- with specified load profile,
- under effects of fire protection actuation, and
- with specified fuel type and quality, combustion air contaminants auxiliary, service water quality, and electrical power supply requirements.

EDG-Specific Design Requirements: Starting and Loading

The EDG should be designed such that at any time during the loading sequence frequency and voltage transients will be within the following limits:

- frequency will not decrease to less than 95% of nominal,
- frequency should be restored to within 2% of nominal in less than 60% of each load-sequence interval for a stepload increase, and less than 80% of each load-sequence interval for disconnection of the single largest load,
- upon the largest load rejection EDG speed should not exceed the nominal speed plus 75% of the difference between nominal speed and the overspeed trip set point, or 115% of nominal (whichever is lower),
- upon 100% load rejection EDG speed should not reach the overspeed trip set point,
- voltage will not decrease to less than 75% of nominal, and
- voltage should be restored to within 10% of nominal within 60% of each load-sequence interval.

A larger decrease in voltage and frequency may be justified for a diesel generator that carries only one large connected load, as there would be no other loads that may be affected by the voltage and frequency decrease.

The load sequence time interval must be justified by analysis, which will prove that the interval is adequate to allow for acceleration of motors without incurring an unacceptable voltage drop. It is generally 5 s, and it should include sufficient margin for the accuracy and repeatability of the load-sequence control logic.

The unit should be capable of starting, accelerating, and being loaded with the design load within the time required. The following conditions should be met:

1. EDG starting from the normal standby condition.
2. EDG starting without cooling, for the time required to bring the cooling equipment into service, when the cooling equipment is powered from the EDG,
3. EDG restarting, with an initial engine temperature equal to the continuous rating engine-generator temperature,
4. EDG design load. The unit should be capable of carrying the design load for the time required by the safety-related station requirements, and
5. EDG light-load or no-load operation. The unit should be capable of accepting design load following operation at light load or no load for the time required by the equipment specification.
6. EDG power quality. The unit should be capable of maintaining voltage and frequency at the generator terminals within limits that will not degrade the performance of any of the loads, below their minimum requirements, including the duration of transients caused by load application and/or load removal.

The diesel generator units may be utilized to the limit of their power capabilities, as defined by the continuous and short time ratings.

Equipment manufacturers impose limitations on light load operation after a loading cycle of 4 h at 30% or less of the continuous rating, the unit should be operated at a load of at least 50% of the continuous rating for a minimum of 0.5 h.

Harmful torsional vibration stresses should not occur within a range of rated idle speed from 10% above to 10% below and from 5% above to 5% below rated full load speed.

Rotating devices should be designed to withstand the overspeed that results from a load rejection. As a minimum, the generator rotor, exciter rotor (if used), and flywheel should be designed to withstand an overspeed of 25% without damage. Margin should be provided to allow the overspeed device to be set sufficiently high to guarantee that the unit will not trip on load rejection.

When the diesel generator unit is on standby, the governor should automatically be placed in the isochronous mode of operation. Likewise, the voltage regulator should be placed in the nonparallel mode.

Upon receipt of an emergency start-diesel signal, the automatic control system should provide automatic startup and automatic adjustment of speed and voltage to a ready-to-load condition. A start signal should override all other operating modes and return control of the diesel generator unit to the automatic control system. When under maintenance, an emergency start signal should not override manual modes.

Provisions should be made for control both from the control room and also external to the control room in case the control room becomes uninhabitable.

EDG-Specific Design Requirements: Surveillance Systems

The EDG unit should be provided with surveillance systems permitting remote and local alarms and indicating the occurrence of abnormal, See IEEE 627 [22] pretrip, or trip conditions.

Principal surveyed items are

a) unit not running,
b) unit running, not loaded,
c) unit running, loaded, and
d) unit out of service.

The following systems should have sufficient mechanical and electric instrumentation to survey the variables required for successful operation and to generate the abnormal, pretrip, and trip signals required for alarm of such conditions:

- starting system,
- lubricating system,
- fuel system,
- primary and secondary cooling system,
- combustion air system,
- exhaust system,
- governor system, and
- generator excitation and voltage regulation system.

6.4   Factory Qualification

General

Factory qualification includes analysis, testing, and documentation. When the EDG leaves the factory, it should be fully operational and qualified for Class 1E Service. After factory qualification, *acceptance testing* is performed upon receipt of the unit at the plant, to confirm the qualification performed at the factory.

Qualification documentation should be in accordance with IEEE 323 [17], and should include the following:

a) service and environment data,
b) classification table of all components as safety related or nonsafety related,
c) analysis/justification for non-safety-related components,
d) classification table of safety-related components, including aging and methods,
e) test data for generator, engine, control panels, exciter/regulator, and other large and small components including seismic qualification, and
f) historical data/justification for items requiring periodic replacement.

The next step in the qualification process is to categorize all safety-related components as components with or without age-related failure mechanisms. Components with age-related failure mechanisms are required to be tested or analyzed to determine their respective qualified life. Components that are in the nonaging category require justification to show why they would not age under the circumstances.

Qualification by testing of the various components is preferred. For the control panels, the individual applicable components are aged, assembled in the unit or on a fixture (simulated mounting), and seismically tested. Documentation of posttest operational parameters is required. Burn-in of a unit with electronic components may be a part of the qualification process. For the generator, aging testing of winding materials is performed followed by seismic simulation. The generator assembly, overall, is seismically analyzed according to IEEE 344. For mechanical components, seismic analysis is accomplished where testing is impractical (e.g., air receivers, engines, and smaller mechanical off-engine components such as pumps, filters, strainers).

Complete documentation of all seismic test and analysis is required. Radiation testing is not required; a brief material analysis is sufficient.

Factory Qualification: Testing and Analyses

Each EDG unit should be tested at the factory, while testing is preferred, analyses may supplement test or be substituted for test, where testing is not practical. Analysis should include the basis of assumptions used.

Certain components, such as the exhaust muffler, intake air filter silencer, and radiator may be substituted in place of the equipment to be provided for a specific site, since it is not practical to utilize the equipment and piping that will exist at the. However, results of tests should be corrected to the conditions of the final site.

Factory Qualification: Engine Tests

Each engine should be tested, utilizing either a water brake dynamometer or a generator to provide accurate means to control power delivery. The following tests should be performed:

a) break-in test and
b) performance test.

The performance test should be run after the engine has satisfactorily completed the break-in test. The following data should be logged:

    i) speed (rev/min),
    ii) brake (HP or kW),
    iii) brake-specific fuel consumption,
    iv) intake manifold pressure and temperature,
    v) individual cylinder exhaust temperature,
    vi) turbocharger exhaust temperature,
    vii) jacket water temperature and pressure (engine inlet and outlet),
    viii) lube oil temperature (engine inlet and outlet),
    ix) lube oil header pressure (engine inlet),
    x) type of fuel and fuel heat content,
    xi) barometric pressure and intake air temperature, and
    xii) engine-mounted devices, controls, and alarms.

Factory Qualification: Generator Tests

Generator testing should be in accordance with NEMA MG 1: factory production tests of excitation, control, and other accessories/auxiliaries.

Factory Qualification: Initial Type Tests of the EDG Set

A type testing program consisting of load capability, start and load acceptance, and margin tests should be applied to diesel generators not previously type tested as standby power sources for nuclear power generating stations.

Type tests should be performed following successful completion of the factory qualification tests. Following the successful completion of these type tests, the equipment should be inspected in accordance with the manufacturer's standard procedure, and inspection results should be documented.

Load capability tests demonstrate the capability of the diesel generator unit to carry the following rated loads at rated power factor for the period of time indicated, and to successfully reject load:

Test Step 1. Apply load equal to the continuous rating until reaching engine equilibrium temperature.

Test Step 2. Apply short time rated load, immediately following step 1, for a period of 2 h.

Test Step 3. Apply the continuous rated load for a period of 22 h.

Test Step 4. Load rejection test, deemed acceptable if the increase in speed of the diesel engine does not exceed 75% of the difference between nominal speed and the overspeed trip set point, or 15% above nominal, whichever is lower.

Test Step 5. Light-load or no-load capability should be demonstrated. Light-load or no-load operation should be followed by a load application ≥ 50% of the continuous kilowatt rating for a minimum of 0.5 h.

Factory Qualification: Start and Load Acceptance Tests

To demonstrate the EDG reliability, a series of tests should be conducted to establish the capability of the diesel generator unit to start and accept load within the period of time necessary to satisfy the plant design requirement. IEEE 587 provides an accepted test program. This test program requires a total of 100 valid start and load tests with no failures allowed is as follows:

1. Engine cranking should begin upon receipt of the start-diesel signal, and the diesel generator unit should accelerate to specified frequency and voltage within the required time interval.
2. Immediately following step (a), the diesel generator unit should accept a single-step load ≥ 50% of the continuous kilowatt rating. Load may be totally resistive or a combination of resistive and inductive loads.
3. At least 90 of these tests should be performed with the diesel generator unit initially at warm standby, based on jacket water and lube oil temperatures at or below values recommended by the engine manufacturer. After load is applied, the diesel generator unit should continue to operate until jacket water and lube oil temperatures are within ± 5.5°C (± 10°F) of the normal engine operating temperatures for the corresponding load.
4. At least 10 tests should be performed with the engine initially at normal operating temperature equilibrium defined as jacket water and lube oil temperatures within ± 5.5°C (± 10°F) of normal operating temperatures, as established by the engine manufacturer for the corresponding load.

Margin Tests in Factory Qualification

At least two margin tests should be performed using either the same or different load arrangement. A margin test load at least 10% greater than the magnitude of the most severe single-step load within the load profile is considered sufficient for the margin test.

The frequency and voltage excursions recorded may exceed those values specified for the plant design load. The criteria for margin tests are as follows:

1. Demonstrate the ability of the generator and excitation system to accept the margin test load (usually the low power factor, high inrush, and high starting

current of a pump motor) without experiencing instability resulting in generator voltage collapse, or significant evidence of the inability of the voltage to recover.
2. Demonstrate that there is sufficient engine torque available to prevent engine stall and to permit the engine speed to recover, when experiencing the margin test load.

Factory Qualification: Aging Test

Components and assemblies should be classified into one of the two following categories:

a) Components and assemblies required to enable the diesel generator unit to meet its safety-related function. These components and assemblies require consideration of aging as a potential cause for common mode failures. For example, the governor, generator, cable, excitation system, engine, starting air solenoid valves, and those gaskets and seals that are applied to prevent leakage that degrades unit performance. (See 117 [2], IEEE 259 [3], and IEEE 275 [4].)

b) Components and assemblies not required to enable the diesel generator unit to perform a safety-related function. Each of these nonsafety-related components or assemblies requires verification that they will not degrade the safety-related function, which may be accomplished by testing or analysis. For example, generator resistance temperature detectors, neutral grounding equipment, space heaters, starting air compressors and drives, keep-warm heaters and pumps, and those gaskets and seals whose failure will not degrade unit performance.

Components without significant age-related failure mechanisms may be excluded. For example, it is recognized that cast iron used in the basic engine block will not represent a potential age-related failure mechanism over normal nuclear service life.

Following classification, those components with potential age–related failures should be qualified by testing (the preferred method), analysis, or a combination of test and analysis. Components with a resultant identified qualified life less than the overall qualified life objective should have a maintenance/replacement interval defined. If aging by test is used, it should be followed by seismic qualification to meet IEEE Std 344. In all cases, the documentation requirements IEEE Std 323 should be satisfied.

Factory Qualification: Seismic Requirements

Seismic qualification in accordance with IEEE 344 [6] is required for all safety-related components. Nonsafety-related components require analysis or test to show that they will not degrade the safety-related function of the unit during a seismic event. Seismic testing should be followed by functional testing to verify the applicable capabilities.

6.5 Site Acceptance Testing

After final assembly and preliminary startup testing, each EDG unit should be tested at the site. The site acceptance and preoperational testing should include

- starting,
- load acceptance,
- rated load,
- load rejection,
- subsystems, including electrical,
- reliability,
- LOOP, and
- safety injection actuation signal (SIAS).

6.6 Site Preoperational Testing

Reliability tests should be performed to demonstrate that an acceptable level of reliability has been achieved to place the new diesel generators into operation. This should be achieved by a minimum of 25 valid start and load tests without failure on each installed diesel generator.

Preoperational testing should include

- reliability,
- LOOP,
- SIAS,
- combined SIAS and LOOP,
- synchronizing,
- hot restart,
- endurance load, and
- load rejection.

6.7 Site Operational Testing

Operational tests are those performed after the unit has been put in service. They should include the following:

Monthly	Start, load run
Every plant refueling outage	Availability, combined SIAS and LOOP, load rejection, endurance load, hot restart, synchronizing, protective trip bypass, test mode override
Every 6 months	Start, load run
Every 10 years	Independence

Site Periodic Testing Applicable Terms

Start demands: All valid and inadvertent start demands, including all start-only demands and all start demands that are followed by load-run demands, whether by automatic or manual initiation, are start demands. In a start-only demand, the EDG is started, but no attempt is made to load the EDG (see the exceptions below).

Start failures: Any failure within the EDG system that prevents the generator from achieving a specified frequency (or speed) and voltage within specified time allowance is classified as a valid start failure. (For monthly surveillance tests, the EDG can be brought to rated speed and voltage in the time recommended by the manufacturer to minimize stress and wear.) Any condition identified during maintenance inspections (with the EDG in the standby mode) that would definitely have resulted in a start failure if a demand had occurred should count as a valid start demand and failure.

Load-run demands: To be valid, the load-run attempt should follow a successful start and meet one of the following criteria (see the exceptions below):

a) load-run of any duration that results from a real (i.e., not a test) automatic or manual signal and
b) load-run test to satisfy the plant's load and duration test specifications other operations (e.g., special tests) in which the EDG is planned to run for at least 1 h with at least 50% of design load

Load-run failures: A load-run failure should be counted when the EDG starts but does not pick up the load and run successfully. Any failure during a valid load-run demand should count (see the exceptions below). (For monthly surveillance tests, the EDG can be loaded at the rate recommended by the manufacturer to minimize stress and wear.) Any condition identified during maintenance inspections (with the EDG in the standby mode) that definitely would have resulted in a load-run failure if a demand had occurred should count as a valid load-run demand and failure.

Exceptions: Unsuccessful attempts to start or load-run should not count as valid demands or failures when they can definitely be attributed to any of the following:

- any operation of a trip that would be bypassed in the emergency operation mode (e.g., high cooling water temperature trip),
- malfunction of equipment that is not required to operate during the emergency operating mode (e.g., synchronizing circuitry),
- intentional termination of the test because of alarmed or observed abnormal conditions (e.g., small water or oil leaks) that would not have ultimately resulted in significant damage or failure of the emergency generator,
- component malfunctions or operating errors that did not prevent the EDG from being restarted and brought to load within 5 min (i.e., without corrective maintenance or significant problem diagnosis), and
- a failure to start because a portion of the starting system was disabled for test purposes, if followed by a successful start with the starting system in its normal alignment Each diesel generator valid failure that results in declaration of the EDG as being inoperable should count as one demand and one failure. Exploratory tests during corrective or preventive maintenance should not count as demands or failures. However, the successful test that is performed to declare the EDG operable should count as a demand.

Site Periodic Testing

After being placed in service, the diesel generator unit needs to be tested periodically to demonstrate the unit's continued capability and availability to perform its intended function.

Records and analysis should be maintained for all periodic tests. It is not necessary to begin these tests from standby conditions unless otherwise specified. Applicable test considerations are

a) Test equipment should not cause a loss of independence between redundant diesel generator units or between diesel generator load groups.
b) Periodic testing of a diesel generator unit should not impair the capability of the unit to supply emergency power in the required time in response to start- diesel signals.
c) All diesel generator unit protective trips and alarms should be operative during applicable periodic testing.
d) Written procedures for testing should be prepared and utilized. The procedures should include the manufacturer's applicable test recommendations and should identify all special arrangements or changes in the normal system

configuration required to perform the test. The procedures should ensure that the system is restored to its normal configuration after completion of the tests.

All tests should be in general accordance with the manufacturer's recommendations for reducing engine wear, including cooling down operation at reduced power followed by postoperation lubrication. (See IEEE 383 [8], and IEEE 650 [9].)

Periodic Tests

Periodic tests should consist of availability, system operation, and independence verification tests (see IEEE 338 [18]). These tests should be preceded by a prelube period and should be in general accordance with the manufacturer's recommendations for reducing engine wear, including cooling down operation at reduced power followed by postoperation lubrication. Unless otherwise noted, these tests should be performed at a power factor as close as practical to the design load power factor as plant voltage conditions permit.

Detailed procedures for each test should identify special arrangements or changes in normal system configuration that must be made to put the EDG under test. Jumpers and other nonstandard configurations or arrangements should not be used after initial equipment startup testing.

Periodic testing should include (1) endurance and load test, (2) hot restart test, (3) synchronizing test, (refer to IEEE 387 [11] and IEEE 338 for test details), (4) protective-trip bypass test, (5) test mode override, (6) independence, (7) test mode override test, (8) slow-start test, (9) rated load test, (10) fast-start test, (11) loss of off-site power test, (12) SIAS test, (13) combined SIAS and LOOP test, (14) largest load rejection test, (15) design-load rejection test, (16) endurance and load test, (17) hot restart test, (18) test mode override test, (19) electrical test, (20) subsystem test, (21) availability tests, (22) system operation tests, and (23) ongoing surveillance site testing.

Periodic Monthly Testing

After completion of the reliability demonstration during preoperational testing, the EDGs should be periodically tested during normal plant operation. Each diesel generator should be started and loaded at least once every 31 days (with the maximum allowable extension not to exceed 25% of the surveillance interval).

Periodic 6-Month (or 184-Day) Testing

This test may substitute for a monthly test. To demonstrate the capability of the EDG to start from standby conditions and provide the necessary power to mitigate a loss-of-coolant accident (LOCA) coincident with a LOOP, each diesel generator should be started from standby conditions once every 6 months as described in IEEE 387. This will verify that the diesel generator reaches the required voltage and frequency within acceptable limits and time as specified in the plant technical specifications. Following this test, the EDG should be loaded.

Periodic Testing at Refueling Outage

The capability of the overall EDG design should be demonstrated during every refueling outage (or at a frequency of not more than every 24 months) by performing the tests specified. Certain tests may be conducted during the operating mode with NRC approval if the tests can be safely performed without increasing the probability of plant trip, loss of power to the safety buses, or LOOP.

Periodic 10-Year Testing

This testing involves demonstrating that the trains of standby electric power are independent at a frequency of once every 10 years (during a plant shutdown) or after any modifications that could affect EDG independence (whichever is shorter) by starting all redundant units simultaneously to identify certain common failure modes undetected in single diesel generator unit tests.

Test Loads

Loads to be applied, carried, and rejected during site testing should be the design load auxiliaries located at the station. Equivalent loads may be used if these auxiliaries cannot be operated for testing.

Reporting Criteria

Plant owners must conform to the reporting requirements of 10 CFR Part 21, "Reporting of Defects and Noncompliance" [2]; 10 CFR 50.72, "Immediate Notification Requirements for Operating Nuclear Power Reactors"; and 10 CFR 50.73, "License Event Reporting System."

Ongoing surveillance may be used as a basis for the identification of equipment degradation and validation of the results of the aging and aged equipment testing.

EDGs should be designed so that they can be tested to simulate the parameters of operation (e.g., manual start, automatic start, load sequencing, load shedding, operation time), normal standby conditions, and environments (e.g., temperature, humidity) that would be expected if actual demands were placed on the system. If prelubrication systems or prewarming systems designed to maintain lube oil and jacket water cooling at certain temperatures (or both) are normally in operation, this would constitute normal standby conditions for the given plant.

Plant design provisions should include the capability to test each EDG independently of the redundant units. Test equipment should not cause a loss of independence between redundant diesel generators or between diesel generator load groups. Testability should be considered in selecting and locating instrumentation sensors and critical components (e.g., governor, starting system components). Instrumentation sensors should be readily accessible and designed so that their inspection and calibration can be verified in place. The plant design should include status indication and alarm features.

Test Parameters to be Recorded

Parameters to be recorded during tests should include the following:

- lube oil system pressures and temperatures,
- lube oil levels,
- Electrical: frequency, power, reactive power, generator current and voltage, field excitation current,
- engine cooling water pressure and temperature,
- lube oil cooler water pressure and temperature,
- ambient air temperature, barometric pressure, and relative humidity,
- crankcase vacuum or pressure,
- amount of oil added,
- lubricating oil analysis,
- engine cooling analysis, and
- fuel pump rack settings.

Records

Records should be maintained for each diesel generator unit to provide a basis for an analysis of the unit's overall performance. The records should be retrievable and should provide a basis for verifying any assumptions made. The documentation may also be used to shorten or extend the replacement intervals or to extend equipment or station life. The records should include, as a minimum, the following features:

a) all start attempts, including those from bona fide signals; maintenance, repair, and out-of-service time histories, as well as cumulative maintenance and operating data; cumulative statistical analysis of diesel generator unit test results, together with results of operation of the diesel generator unit when required by actual demand;

b) critical failure mechanisms, human errors, and common mode failures, including cause and corrective action; and

c) test parameter data that may have application for reliability and data trending.

6.8   Site Periodic Testing and Surveillance: Preventive Maintenance Program

A separate preventive maintenance, inspection, and testing program should be established for the diesel generator unit and all supporting systems based on the manufacturer's recommendations, including time intervals for parts replacement of those components with a qualified life less than the unit qualified life objective. These recommendations may be based on operating hours or fixed time intervals, or both.

Procedures should include, as a minimum, specific programs for each portion of the unit as follows:

a) the engine, including the governor, overspeed trip device, turbocharger, lube oil components, fuel oil components, cooling water components, water chemistry, starting components, intake air, exhaust, and crankcase ventilation;
b) the generator and exciter, including insulation, bearings, cooling system, lubricating system, and space heaters;
c) electrical auxiliary equipment, including local engine and generator control, exciter and voltage regulator, and protection and surveillance components.

Records and Analysis

Records should be maintained for each diesel generator unit to provide a basis for an analysis of the unit's overall performance. These records also provide a basis for verifying any assumptions made concerning age-related failure mechanisms. The documentation may also be used to shorten or extend the replacement intervals.

Modifications

Modifications to a previously qualified diesel generator unit, such as governor, generator, overall system flywheel effect, excitation system characteristics, cooling media, or other accessories/auxiliaries that may change the capability or performance of a previously qualified diesel generator unit, should be analyzed to determine if the degree of change is major or minor.

a) Major changes to a qualified engine, such as changes in stroke or bore, brake mean effective pressure, speed, or diesel generator arrangement in unique or different configuration should be requalified.
b) Minor changes to a qualified diesel generator unit, such as component parts substitution, should be qualified by analysis or testing, or both.

Recommended Program for EDG Monitoring and Trending Parameters

The industry and regulatory philosophy is shifting away from reliance on statistical testing and toward monitoring and trending to facilitate aging/reliability determinations. The required data may be obtained by monthly testing. The test duration should be adequate to obtain several sets of data to confirm engine and generator operability.

The purpose of the monthly tests is to perform a "health check" of the engine and generator system. If most of the recommended parameters are checked each month, immediate engine condition is determined. By trending certain parameters, the long-term degradation mechanisms can also be determined. These parameters may be recorded each month for unit management purposes.

Recommended Parameters

Many installations have the necessary sensors and gages already in place to obtain this data. Some installations have fewer sensors and gages installed. Where additional sensors and gages are needed, commercial grade components may be installed since the data is advisory in nature. These commercial-grade components would not jeopardize the unit safety function if they fail to operate.

The regular review and use of these recommended parameters should increase the unit reliability and reduce aging concerns by detecting substandard performance early before failures occur. Monitoring and trending use in diesel generator unit management is now generally considered more practical for reliability assurance than depending upon start and run statistical information.

Diesel Generator Unit Reliability Program Elements

Nuclear power generating stations are developing formal reliability programs for the installed diesel generator units. Guidelines for an effective program have been developed and published by the Nuclear Management and Resources Council (NUMARC), now known as the Nuclear Energy Institute (NEI). This description is intended to highlight the technical points and major diesel generator unit reliability program elements that should be universally applicable. Station-specific administrative details and supporting formal procedures should be developed to amplify these major reliability elements.

Major Reliability Program Elements

The principal elements of an effective diesel generator unit reliability program are the following:

a) overall diesel generator unit reliability program plan,
b) defined management and technical responsibilities for each program element,

c) plant-specific surveillance and monitoring plan,
d) plant record and data management system designed for easy data retrieval,
e) defined maintenance program with a reliability focus,
f) procedures for failure identification, root cause analysis, and future failure prevention, and
g) procedures and criteria for problem closeout and follow-up.

The principal element of a reliability program is an overall plan. The plan should specifically address all other plan elements by addressing their appropriate importance in an integrated manner. Plan procedures, surveillance, training, monitoring and trending, and failure resolution documentation should all be a part of the reliability plan. The management and technical responsibilities for each program element should be established in written procedures. The plan should address training of qualified personnel in detailed engine and governor maintenance procedures similar to that offered by manufacturers of this equipment. An overall responsibility for diesel generator unit related matters and coordination may give the best results.

The periodic testing and monitoring element should address reliability by use of monthly system operation intended to determine the condition of important engine/generator parameters. These typically should consist of oil, water, and gas temperatures and pressures in the various engine subsystems during operation and while generating power equal to the plant emergency power requirements. Daily, weekly, monthly, and quarterly surveillance should also be specified. Engine and generator operating parameters and surveillances associated with degraded performance and aging should be trended, where such trending could detect incipient failures and permit corrective maintenance before the actual failure occurs.

A record and data management system, designed for easy data retrieval, should be available for use by plant personnel. Diesel generator unit records are important and should support the other program elements. Measures should be taken to safely store these records and prevent their loss. A computer-managed system is recommended to reduce costs and improve access to the data.

The maintenance program should have a reliability improvement intent. In some cases, past practices of certain preventive maintenance procedures and routine disassembly for inspection purposes have been shown to have negative results on reliability. A good maintenance program should, over time, eliminate those procedures. Systems with higher failure rates, such as the instrument and control system, should receive additional maintenance effort for improved reliability. Diesel generator unit system modifications may be included in the plan under the maintenance responsibilities. Reliability considerations should be a part of the modification review process.

Failure identification, analysis for failure root cause, and correction should be defined in procedures. Each analysis should have the intent to reduce failures by identifying a corrective action to prevent future failures. For some failures of an intermittent nature, it may be impossible to assign an exact root cause.

Problem closeout should include follow-up to ensure that the problem has been corrected and that related problems are not recurring. Failures with a recurring root cause show that corrective actions taken in the past were not correct and a new analysis should be performed with a more appropriate corrective action as the expected result. The plant records should be reviewed for each analysis and problem closeout to ensure that recurring failures are eliminated.

Questions and Problems

6.1 EDG governor

The automatic speed control regulation band of an EDG allows for frequency excursions in the band of ± 3% during operation at accident conditions. Calculate the impact on the EDG loading for an excursion to the maximum expected limits, assuming the EDG is loaded to 100% with the machine running at 97% speed, and if load consists of 90% of motors driving centrifugal pumps and 10% of resistance-type loads.

6.2 EDG Frequency variation

Determine the EDG kVA and kW ratings for a nuclear plant, assuming available generators from the industry are rated 0.80 or 0.85 power factor. The EDG should operate successfully for the following conditions:

- Generator frequency band 0.97–1.03 pu on 60 Hz base
- Generator terminal voltage 0.95–1.05 pu

The plant load under accident conditions is 3000 kVA at 0.9 power factor, nominal 100% frequency, and nominal voltage of 4.16 kV. The load is made up of 80% of motors driving centrifugal pumps (constant kVA) and 20% static loads (constant impedance).

The EDG should be capable of driving the accident load when its terminal voltage at 105% of nominal.

6.3 EDG Rating

A nuclear plant EDG is rated at 3000 kVA, 0.85 power factor. The plant operation under a LOOP for 3 days which requires the EDG to be loaded at 3000 kVA, 0.85 power factor, on a continuous basis. The driving diesel engine has a maximum rating of 2400 kW on a continuous basis and a 2-h rating of 2660 kW. Evaluate the adequacy of the EDG to cope with the LOOP that lasts for 3 days.

6.4   Fire Suppression

An EDG has a water fire suppression system, what precautions are necessary to prevent damage to the generator in case of fire suppression system activation?

6.5   EDG Loading

An EDG loads are 20% constant impedance. 70% constant kVA, 10% constant current.

Assume the voltage regulator malfunctions and the voltage goes up to 110%. What will be the resultant percent load on the EDG?

6.6   EDG monthly test

An EDG is synchronized to the grid for the monthly testing, when the grid voltage suddenly decays to 50%.
- Provide the effect on the operation of the plant and on the EDG.
- Name the type of protection available for the EDG and explain its operation.

6.7   An EDG has a neutral ground protection consisting of an overvoltage relay connected across the grounding resistor (this is an over-simplified description of the ground fault protection). Upon the presence of phase to ground fault, the relay would be actuated by the voltage developed by the passage of the ground fault current. The plant was licensed with the approach of bypassing protection which is not considered essential under DBE conditions and has included a design to bypass the EDG ground fault protection, but not the generator differential protection. Assume the plant is mitigating a design bases event. Will the generator be automatically tripped?

Provide justification for your answer.
- **a)** Will the generator be automatically tripped by the differential protection?
- **b)** Will the generator be automatically tripped by the ground fault protection?
- **c)** Will there just be an alarm in the control room?
- **d)** If the answer to (c) is affirmative, what will the operator do?

References

1 US Code of Federal Regulations, Title 10, Part 50, "Domestic Licensing of Production and Utilization Facilities," US Nuclear Regulatory Commission.
2 IEEE Std 117, "IEEE Standard Test Procedure for Evaluation of Systems of Insulating Materials for Random-Wound AC Electric Machinery (ANSI)."

3 IEEE Std 259, "IEEE Standard Test Procedure for Evaluation of Systems of Insulation for Specialty Transformers (ANSI)."

4 IEEE Std 275, "IEEE Recommended Practice for Thermal Evaluation of Insulation Systems for Alternating-Current Electric Machinery Employing Form-Wound Preinsulated Stator Coils for Machines Rated 6900 V and Below (ANSI)."

5 IEEE Std 334, "IEEE Standard for Qualifying Continuous Duty Class 1E Motors for Nuclear Power Generating Stations (ANSI)."

6 IEEE Std 344, "IEEE Recommended Practice for Seismic Qualification of Class 1E Equipment for Nuclear Power Generating Stations (ANSI)."

7 IEEE Std 382, "IEEE Standard for Qualification of Actuators for Power Operated Valve Assemblies with Safety-Related Functions for Nuclear Power Plants (ANSI)."

8 IEEE Std 383, "IEEE Standard for Type Test of Class 1E Electric Cables, Field Splices, and Connections for Nuclear Power Generating Stations (ANSI)".

9 IEEE Std 650, "IEEE Standard for Qualification of Class 1E Static Battery Chargers and Inverters for Nuclear Power Generating Stations (ANSI)."

10 US Code of Federal Regulations, Title 10, Part 21, "Reporting of Defects and Noncompliance," US Nuclear Regulatory Commission, Washington, DC.

11 IEEE Std 387, "IEEE Standard Criteria for Diesel-Generator Units Applied as Standby Power Supplies for Nuclear Power Generating Stations."

12 ANS 59.51, "Fuel Oil Systems for Emergency Diesel Generators."

13 ANSI C50.10, "American National Standard General Requirements for Synchronous Machines."

14 ANSI/ASME, "Boiler and Pressure Vessel Code."

15 ANSI/NFPA 37, "Stationary Combustion Engines and Gas Turbines."

16 IEEE Std 308, "IEEE Standard Criteria for Class 1E Power Systems for Nuclear Power Generating Stations (ANSI)."

17 IEEE Std 323, "IEEE Standard for Qualifying Class 1E Equipment for Nuclear Power Generating Stations (ANSI).

18 IEEE Std 338, "IEEE Standard Criteria for the Periodic Surveillance Testing of Nuclear Power Generating Station Safety Systems (ANSI)."

19 Regulatory Guide 1.155, "Station Blackout," US Nuclear Regulatory Commission,Washington, DC.

20 IEEE Std 384, "IEEE Standard Criteria for Independence of Class 1E Equipment and Circuits (ANSI)."

21 IEEE Std 603, "IEEE Standard Criteria for Safety Systems for Nuclear Power Generating Stations (ANSI)."

22 IEEE Std 627, "IEEE Standard for Design Qualification of Safety Equipment Used in Nuclear Power Generating Stations (ANSI)."

23 IEEE Std 741, "IEEE Standard Criteria for the Protection of Class 1E Power Systems and Equipment in Nuclear Power Generating Stations (ANSI)."

7

On-Site Emergency Direct Current Source

7.1 Energy Storage Systems for Nuclear Generating Stations

The most commonly utilized energy storage system for nuclear generating stations is direct current (DC) batteries, based on the electrochemical principle of electricity storage (Figure 7.1). These systems have been in use at fossil power plants, electrical substations, and other industry for many years and have provided highly reliable service. Of the types of batteries in use, the most popular is the lead acid type. In the United States, there is one nuclear generating station that utilizes a hydroelectric plant to help it meet the energy storage requirements. Nevertheless, this plant utilizes batteries as the main DC energy storage.

7.2 General Requirements of Direct Current Systems

The DC power systems include power supplies and distribution systems arranged to provide power to the Class 1E direct current loads comprising the power, control, and of the Class 1E power systems. To meet the single failure criterion, features such as physical separation, electrical isolation, redundancy, and qualified equipment should be included in the design to aid in preventing a mechanism by which a single design basis event can cause redundant equipment within the station's Class 1E power system to be inoperable. The battery system should consist of the storage cells, connectors, and connections to the distribution system supply circuit-interrupting device.

Electrical Systems for Nuclear Power Plants, First Edition. Omar S. Mazzoni.
© 2019 by The Institute of Electrical and Electronic Engineers, Inc. Published 2019 by John Wiley & Sons, Inc.

Figure 7.1 Typical battery installation.

7.3 Design Requirements

Operating Mode of Direct Current Storage Systems

The DC storage systems for nuclear generating stations are designed for operation with a battery charger serving to maintain the battery in a charged condition as well as to supply the normal DC load.

DC systems are normally operated ungrounded, that is, neither the positive nor the negative is connected to ground. The reason for this is that reliability is enhanced by avoiding tripping underground faults. A ground detection system is normally provided to alert the operator when a ground fault occurs.

Capability, Availability, Independence, and Testing of Battery Systems

The battery system should be capable of starting and operating its required steady-state and transient loads. The battery system stored energy should be sufficient to provide an adequate source of power for starting and operating all required connected loads during an interval of time when either of the following occur:

- Alternating current (ac) to the battery charger is lost for the time stated in the design basis, or

- ac to the battery charger has been restored, the battery is being restored to its fully charged state, and power in excess of the capacity of the battery charger is needed.

The battery system should be permanently available during normal operations and immediately ready to supply the loads following the loss of power from the ac system.

Distribution circuits to redundant equipment should be physically and electrically independent of each other in accordance with IEEE 384. No provision should be made for automatically interconnecting redundant load groups. If nonautomatic interconnecting means are furnished, provision should be included that prevents paralleling of the redundant DC sources.

No provision should be made for automatically transferring loads between Class 1E power sources.

Connecting cables between the Class 1E power systems and systems located in nonsafety class structures should be provided with Class 1E automatic circuit-interrupting devices located in a safety class structure.

Provisions should be made to perform battery capacity tests in accordance with IEEE 450 [5]. Qualification of battery systems should be in accordance with IEEE 535 [2], and IEEE 627 [3].

Installation of Battery Systems

DC system Class 1E electric loads should be separated into a minimum of two redundant load groups.

The protective actions of each load group should be independent of the protective actions provided by redundant load groups. Each of the redundant load groups should be connected to a power supply that consists of one or more batteries and one or more battery chargers.

The battery charging system should consist of a battery charger (or chargers) for each load group. Two or more chargers may have a common ac power supply provided that the consequences of the loss of the power supply to the load group under design basis conditions are acceptable.

IEEE 484 provides detail recommendations on installation design and installation practices for batteries.

7.4 Battery Loads

The battery must supply the DC power requirements when either of the following conditions take place:

- There is a deficiency between the charger output and the DC system load demand.
- Battery charger input power is interrupted.

The battery should be sized for the most severe of these conditions, considering battery load and duration.

Classification of Loads in Terms of Service Duration Requirements

The individual DC loads supplied by the battery during the duty cycle may be classified as continuous or noncontinuous. Noncontinuous loads lasting 1 min or less are designated "momentary loads" and should be given special consideration.

Continuous Loads
Energized throughout the duty cycle. These loads are those normally carried by the battery charger and those initiated at the inception of the duty cycle. Typical continuous loads are as follows:

a) lighting,
b) continuously operating motors,
c) converters (e.g., inverters),
d) indicating lights,
e) continuously energized coils,
f) annunciator loads, and
g) communication systems.

Noncontinuous Loads
These loads may run for a set length of time, be removed automatically, or by operator action. Typical noncontinuous loads are as follows:

a) emergency pump motors,
b) critical ventilation system motors,
c) fire protection systems actuations, and
d) motor-driven valve operations (stroke time >1 min).

Momentary Loads
Momentary loads can occur one or more times during the duty cycle but are of short duration, not exceeding 1 min at any occurrence. Although momentary loads may exist for only a fraction of a second, it is common practice to consider each load will last for a full minute because the battery voltage drop after several seconds often determines the battery's 1 min rating. When several momentary loads occur within the same 1 min period and a discrete sequence cannot be established, the load for the 1 min period should be assumed to be the sum of all momentary loads occurring within that minute. If a discrete sequence can be established, the load for the period should be assumed to be the maximum load at any instant. Sizing for a load lasting only a fraction of a second, based on the battery's 1 min performance rating, results in a conservatively sized battery.

Consult the battery manufacturer for ratings of discharge durations less than 1 min.

Typical momentary loads are as follows:

a) breaker operations,
b) motor-driven valve operations (stroke time >1 min),
c) isolating switch operations,
d) field flashing of generators,
e) motor starting currents, and
f) inrush currents.

When evaluating the size of an existing battery by means of a service test, the actual duration of the momentary load and sequence should be duplicated as accurately as practical

7.5 Classification of Loads in Terms of Power versus Voltage Characteristics

Given that a battery is not a constant voltage device (the output voltage decreases as the battery discharges), the battery loads need to be classified in accordance with their voltage dependence, as follows:

- constant power (current increases as the voltage decreases),
- constant resistance (current decreases as the voltage decreases), and
- constant current (current remains constant as voltage decreases).

Each of the connected loads to the battery needs to be carefully reviewed to determine the maximum possible loads connected to the battery.

Converting Loads to Constant Current Loads

For sizing of battery cells, the loading cycles of current versus time assume that in each cycle, the current is remains constant for the time duration of the cycle (see a battery duty cycle below). Therefore, all loads need to be converted to constant current type for the sizing of the battery cells (see IEEE 485).

Battery terminal voltage decreases (as will the voltage at the loads) as the battery discharges. The magnitude by which the battery voltage decreases depends on the internal battery resistance and the magnitude of the load connected to the battery.

For constant power loads, current increases with a voltage decrease. Inverters and DC/DC power supplies are usually constant power type, as they are internally regulated to maintain a constant output voltage as the input voltage magnitude changes. As a result, the DC input current to the load will increase when its input voltage decreases. If the constant power load is somewhat remote from the battery, the voltage drop may be increased by the cable resistance and the resulting input current will be even higher.

For constant power loads, the following equation applies for the increase in load current as battery voltage declines:

$$I_{AVG} = P/E_{AVG} \qquad (7.1)$$

where I_{AVG} is the average discharge current (in amperes) for the discharge period, P is the discharge load (in watts), and E_{AVG} is the average discharge voltage for the discharge period.

Since the average battery voltage is dependent on the type of cell design and load duration, and since it is often unknown without information from the manufacturer, a conservative method (in terms of load estimation) of converting watts to amperes assumes a constant *current* for the entire load duration as equal to the current being supplied by the battery at the end of the discharge period (minimum volts, maximum amperes). Thus

$$I_{max} = P/E_{min \text{ (at load terminals)}} \qquad (7.2)$$

where I_{max} is the discharge current at the end of the discharge period and P is the discharge power in watts.

Example: For a 24-cell battery operating in a nominal 48 V system with a minimum battery voltage of 42 V and a voltage drop from the battery to the load of 2 V, a constant power load of 5000 W will discharge the battery at a rate no greater than

$$5000 \text{ W}/40 = I_{MAX} = 125 \text{ A at 40 V.}$$

For constant resistance loads, current decreases as the voltage decreases. DC motor starting, emergency lighting, relays, contactors, and indicating lights are usually constant resistance. A constant resistance load may be conservatively estimated as a constant current load as follows:

$$I_{MAX} = (W_R)/E_{OC} = (E_{OC})/(R_{AVG}) \qquad (7.3)$$

$$E_{OC} = W_R/I_{MAX} \qquad (7.4)$$

where E_{OC} is battery open circuit voltage, R_{AVG} is an average resistance of the load, and W_R is rated power value.

Similarly, as for constant power loads, the load current can be calculated using the average battery voltage. System voltage drop to the loads can also be included.

However, if the battery requires significant motor starts at the beginning of the cycle, the battery voltage may be calculated from initial data using an estimate of the rated motor starting current, and then checking that the initial voltage will support that level of current, iterating the level of current and voltage until a satisfactory solution is obtained.

For constant current loads, current is approximately constant as the voltage decreases. Running DC motors can be approximated as constant current. Within the normal battery voltage range, the flux is fairly constant in the motor. Modeling a DC motor as a constant current load is conservative if the voltage maintains the motor in saturation.

Battery Duty Cycle Diagram

A battery duty cycle diagram showing the total load at any time during the battery discharge cycle is an aid in the analysis of the duty cycle. To prepare such a diagram, all loads (expressed in either current or power) expected during the cycle are tabulated along with their anticipated inception and shutdown times. The total time span of the duty cycle is determined by the requirements of the nuclear plant emergency services.

Loads are plotted on the diagram as they would occur, with inception and shutdown times. If the inception time is known, but the shutdown time is undefined, it should be assumed that the load will continue through the remainder of the duty cycle.

Random Loads

For loads that occur at random, they should be included at the most critical time of the duty cycle, to simulate the worst-case load on the battery. These loads may be noncontinuous or momentary loads. To determine the most critical time, it is necessary to size the battery without the random load(s) and to identify the section of the duty cycle that controls battery size. Then the random load(s) should be superimposed on the end of that controlling section of the duty cycle.

For an example of a typical battery duty cycle, see IEEE 485.

Cell Selection

This section summarizes some factors that should be considered in selecting a cell design for a particular application. Various cell designs have different charge, discharge, and aging characteristics. IEEE 1184 [8] contains a discussion of cell characteristics.

The following factors should be considered in the selection of the cell:

a) physical characteristics, such as dimensions and weight of the cells, container material, intercell connectors, and terminals;

b) planned life of the installation and expected life of the cell;
c) frequency and depth of discharge;
d) ambient temperature (note that sustained high ambient temperatures result in reduced battery life; see IEEE 484);
e) charging characteristics;
f) maintenance requirements;
g) cell orientation requirements (valve regulated lead acid [VRLA]);
h) ventilation requirements (VRLA);
i) seismic characteristics.

Determining Battery Size

Several basic factors govern the size (the number of cells and rated capacity) of the battery: the maximum system voltage, the minimum system voltage, correction factors, and the duty cycle (see AIEE paper by Hoxie [7]). Since a battery is usually composed of a number of identical cells connected in series, the voltage of the battery is the voltage of a cell multiplied by the number of cells in series.

The ampere-hour capacity of a battery is the same as the ampere-hour capacity of a single cell, which depends upon the dimensions and number of plates.

If cells of sufficiently large capacity are not available, then two or more strings (equal numbers of series-connected cells) may be connected in parallel to obtain the necessary capacity. The capacity of such a battery is the sum of the capacities of the strings. The battery manufacturer may impose limitations on paralleling.

Operating conditions can change the available capacity of the battery, for example:

a) The available capacity of the battery decreases as its temperature decreases.
b) The available capacity decreases as the discharge rate increases.
c) The minimum specified cell voltage at any time during the battery discharge cycle limits the available capacity of the battery.

Number of Cells

The maximum and minimum allowable system voltage determines the number of cells in the battery. It has been common practice to use 12, 24, 60, or 120 cells for nominal system voltages of 24, 48, 125, or 250 V, respectively. In some cases, it may be desirable to vary from this practice to more closely match the battery to system voltage limitations. It should be noted that the use of the widest possible voltage window, within the confines of individual load requirements, will result in the most economical battery. Furthermore, the use of the largest number of cells allows the lowest minimum cell voltage and, therefore, the smallest size cell for the duty cycle.

Calculation of Number of Cells and Minimum Cell Voltage

When the battery voltage is not allowed to exceed a given maximum system voltage, the number of cells will be limited by the cell voltage required for satisfactory charging or equalizing. That is,

Maximum system voltage

$$= (\text{Number of cells}) \times (\text{Cell voltage required for charging})$$

The minimum battery voltage equals the minimum system voltage plus cable voltage drop. The minimum battery voltage is then used to calculate the allowable minimum cell voltage.

$$\text{Minimum battery voltage} = \text{Minimum cell voltage} \times \text{Number of cells}$$

Float Voltage as Limiting Factor

To eliminate the need for frequent equalizing charges (refer to IEEE 450), it may be desirable to establish a float voltage at the high end of the manufacturer's recommended range. The float voltage must, however, be consistent with the maximum system voltage. This higher float voltage may then reduce the number of cells and may increase the cell size required for a given battery duty cycle.

If the calculations indicate a need for a fractional cell, the result should be rounded off to a whole number of cells. The minimum cell voltage, float voltage, and equalize voltage should then be recalculated and verified for adequacy of operation.

Temperature Correction Factor

The available capacity of a cell is affected by its operating temperature. The standard temperature for rating a cell capacity is 25°C (77°F). If the lowest expected electrolyte temperature is below this standard temperature, the cell to be selected should have a capacity available at the lowest expected temperature. If the lowest expected electrolyte temperature is above 25°C (77°F), it is a conservative practice to select a cell size to match the required capacity at the standard temperature and to recognize the resulting increase in available capacity as part of the overall design margin.

Design Margin

It is prudent to provide a capacity margin to allow for unforeseen additions to the DC system and less than optimum operating conditions of the battery due to improper maintenance, recent discharge, or ambient temperatures lower than anticipated, or a combination of these factors. A method of providing this design margin is to add 10–15% to the cell size determined by calculations. If the various loads are expected to grow at different rates, it may be more accurate to apply the expected growth rate to each load for a given time and to develop a duty cycle from the results.

The cell size calculated for a specific application will seldom match a commercially available cell exactly, and it is normal procedure to select the next higher capacity cell. The additional capacity obtained can be considered part of the design margin.

Note that the "margins" required by 6.3.1.5 and 6.3.3 of IEEE 323 [1] are to be applied during "qualification" and are not related to the "design margin" included in this section.

Aging Factor

As a rule, the performance of a lead acid battery is relatively stable throughout most of its life, but begins to decline with increasing rapidity in its latter stages, with the "knee" of its life versus performance curve occurring at approximately 80% of its rated performance.

IEEE 450 recommends that a battery be replaced when its actual performance drops to 80% of its rated performance because there is little life to be gained by allowing operation beyond this point, as battery capacity should decay abruptly beyond 80%. Therefore, to ensure that the battery is capable of meeting its design loads throughout its service life, the battery's rated capacity should be at least 125% (1.25 aging factor) of the load expected at the end of its service life.

Exceptions to this rule exist. For example, some manufacturers recommend that vented batteries with Planté, modified Planté, and round plate designs be their rated capacity (1.00 aging factor). These designs maintain a fairly constant capacity throughout their life.

Initial Capacity

Initial capacity may be less than 100%, because capacity will rise to rated value in normal service after several charge–discharge cycles or after several years of float operation. Nevertheless, for service at nuclear generating stations 100% capacity upon delivery should be specified, as initial capacity can be as low as 90% of rated capacity.

If the designer has provided a 1.25 aging factor, there is no need for the battery to have full rated capacity upon delivery because the capacity normally available from a new battery will be above the duty cycle requirement. When a 1.00 aging factor is used, the designer should ensure that the initial capacity upon delivery is at least 100%, or that there is sufficient margin in the sizing calculation to accommodate a lower initial capacity.

Example: If the cells have 90% initial capacity and the margin is greater than 11%, then no additional compensation for initial capacity is required.

Calculation Method for Cell Size

The calculation method for cell size is described in IEEE 485 [6].

Cell Voltage Profile During Battery Discharge Cycle

Utilizing the battery sizing procedure and methods described in IEEE 485 will ensure that the average cell voltage will not drop below the specified minimum at any point in the duty cycle. However, a method of calculating the voltage at various points in the battery duty cycle, utilizing the battery manufacturer's typical discharge characteristic curves, is also described in IEEE 485.

7.6 Battery Chargers

The battery charger should include all equipment from its connection to the ac system to its distribution system supply circuit breaker connected to the DC bus.

Function, Capability, Availability, Independence, and Testing of Battery Chargers

Each battery charger should furnish electric energy for the steady-state operation of connected loads required during normal and postaccident operation, while maintaining or recharging the battery.

The capacity of each battery charger should be based on the largest combined demands of the various continuous steady-state loads plus charging capacity to restore the battery after the bounding design basis event discharge to a state that the battery can perform its design basis function for subsequent postulated operational and design basis functions. The time period considered for sizing the charger should be as stated in the design basis of the plant. IEEE 946 [4] provides additional recommendations for battery charger sizing.

The battery charger assigned to one load group should be independent of other group battery chargers. Each battery charger should have a disconnecting device in its ac power incoming feeder and its DC power output circuit for isolating the charger.

The battery charger should be designed to prevent the ac power supply from becoming a load on the battery (feedback prevention).

Transients from the ac system should be prevented from affecting the dc system. Likewise, transients in the DC systems should not affect the ac system.

Sizing of Battery Charger

Two equations should be utilized for sizing the battery chargers. The two equations give the output rating of the charger in amperes. The largest current obtained from the two equations is the charger-rated output, as follows:

$$I_1 = I_c + (I_{AH} \times 1.1)/T \qquad (7.5)$$

formula (7.5) is utilized when it is necessary to include the charging time.

$$I_2 = I_c + I_n \qquad (7.6)$$

where I_c is continuous load current of battery, in amperes; I_n is the largest combination of noncontinuous and simultaneous load current of battery under normal plant operation, in amperes; I_{AH} is the ampere-hours discharged from the battery, generally for an overall duration of 8 h; and T is time required to recharge the battery to 95%, in hours, generally 8–12 h.

I_1 or I_2 would be the charger rating depending on the value of $(I_{AH} \times 1.1)/T$, as follows:

- if $(I_{AH} \times 1.1)/T < I_n$, then the charger rating is calculated from equation (7.6).
- if $(I_{AH} \times 1.1)/T > I_n$, then the charger rating is calculated from equation (7.5).

Questions and Problems

7.1 Describe the methodology used to determine battery cell sizing, referring to the applicable IEEE standards as required.

7.2 How is the battery capability periodically verified?

7.3 Describe the method for determining the number of cells of a battery.

7.4 Describe the DC distribution system voltage limitations on the number of cells in a battery.

7.5 Calculate a battery charger rating, given the following data:
- Battery size: 1800 AH based on 8 h rate to 1.75 volts per cell,
- Required recharging time to 95% capacity: 10 h,
- I_n = largest combination of noncontinuous and simultaneous load current of battery under normal plant operation = 150 A, and
- I_c = continuous current = 100 A.

7.6 Provide the methodology for determining remaining life of a battery.

7.7 Assume a nuclear plant running at 100% power and its DC system being connected to a 100% charged, 125 V DC battery and a charger in parallel. Further assume that the DC load current requirement is 66 A.
a) How much current is supplied by the battery to the loads?
b) How much current is supplied by the charger to the loads?

7.8 Provide the reason or reasons for the distribution system to be ungrounded, that is, neither the positive nor the negative is connected to ground.

7.9 Provide the conceptual design of a DC system ground detection, explaining the need for sensitivity of high resistance ground faults.

7.10 Assume a nuclear plant is running at 100% power, and it has a 125-Vdc system that developed a ground fault in the middle of a connected coil. Also assume that the ground detection system operated properly and provided an alarm in the control room. As the DC system is ungrounded, continuous operation with aground fault should be possible. Explain if there is an increase or decrease for the potential of second ground fault to develop.

7.11 Provide the reason(s) for monitoring of the battery room temperature.

7.12 Provide the reason(s) for the continuous exhausting of battery room air, explaining what would be the consequence if the exhaust fan were to fail.

7.13 Determine if the following 125 Vdc load will operate properly for a nuclear plant, under loss of off-site power (LOOP) conditions. Assume the LOOP conditions last for the maximum design duration, that is, those which result in the maximum battery discharge.
Assume:
1. Battery:
  125 Vdc with 58 cells
  Minimum cell voltage: 1.75 V
2. Load
  Minimum operating voltage: 100 Vdc
  Nominal voltage: 120 Vdc
  Type of load: Constant power
  Current requirement at nominal voltage: 3 A
  Distance from the load terminals to the battery terminals: 55 ft
  Feeder size to the load: #12AWG wire,

7.14 Given a 24-cell battery operating in a nominal 48 V system with a minimum battery voltage of 44 V and a voltage drop from the battery to the load of 2 V, a constant power load of 5000 W
What will be the discharge current rate?

7.15 Assume 2.33 V/cell is required for equalize charging of a battery and that the maximum allowable system voltage is (1) 140 V or (2) 135 V.
Calculate the battery number of cells for (a) 140 V maximum allowable voltage and (b) 135 V maximum allowable voltage.

7.16 Given the minimum battery voltage of 105 V, calculate the number of cells, assuming:
a) minimum cell voltage = 1.75 V
b) minimum cell voltage = 1.81 V.

References

1 IEEE 323, "IEEE Standard for Qualifying Class 1E Equipment for Nuclear Power Generating Stations (ANSI)."

2 IEEE 535, "IEEE Standard for Qualification of Class 1E Lead Storage Batteries for Nuclear Power Generating Stations (ANSI)."

3 IEEE 627, "IEEE Standard for Design Qualification of Safety System Equipment Used in Nuclear Power Generating Stations (ANSI)."

4 IEEE 946, "IEEE Recommended Practice for the Design of DC Auxiliary Power Systems for Generating Stations (ANSI)."

5 IEEE 450, "IEEE Recommended Practice for Maintenance, Testing, and Replacement of Vented Lead-Acid Batteries for Stationary Applications."

6 IEEE 485, "Recommended Practice for Sizing Lead-Acid Batteries for Stationary Applications."

7 Hoxie, E. A., "Some Discharge Characteristics of Lead-Acid Batteries," *Transactions of the American Institute of Electrical Engineers, Part II: Applications and Industry*, vol. 73, pp. 17–22, 1954.

8 IEEE 1184, "Guide for the Selection and Sizing of Batteries for Uninteruptible Power Systems."

8

Protective Relaying

8.1 General

This chapter deals with the principal design criteria, design features, and testing requirements for the protection of Class 1E power systems and equipment supplied from those systems. Also reviewed is the protection of those systems not necessarily Class 1E that affect the operation of the Class 1E systems. IEEE Std 741 has been utilized as the guidance document for this chapter.

IEEE and other standards are referenced for additional guidance on specific protection requirements. By definition, the protection should be adequate to sense and to determine the presence of an unacceptable condition and to execute the operations required to limit degradation effects. In this respect, special attention should be given to the requirements for design documentation to support the protection performance.

Included in this chapter is a discussion of the special protection requirements for electrical penetration assemblies (see Chapter 5), installed as part of the containment structure.

Additionally, the following areas of protection are included:

- degraded voltage protection,
- overload protection for valve actuator motor circuits, and
- protection concerns associated with auxiliary system automatic bus transfer.

8.2 General Criteria for the Protection System

The protection should be capable of

a) preventing failures in safety systems and equipment from disabling safety functions to below an acceptable level. The protective actions of each load group should be independent of the protective actions provided by redundant load groups (see IEEE Std 308 [5]);

Electrical Systems for Nuclear Power Plants, First Edition. Omar S. Mazzoni.
© 2019 by The Institute of Electrical and Electronic Engineers, Inc. Published 2019 by John Wiley & Sons, Inc.

b) operating the required devices upon detection of unacceptable conditions to reduce the severity and extent of electrical system disturbances, equipment damage, and potential personnel and property hazards;

c) monitoring the connected preferred power supply and, where an alternate preferred power supply is provided by the design, automatically initiating a transfer or alerting the operator to manually transfer to the preferred alternate power supply;

d) providing indication and identification of the protective operations;

e) periodic testing to verify logic schemes and protective functions;

f) being designed in such a way that the availability of protection control power is monitored;

g) periodic testing to verify set points. This requirement is not applicable to fuses.

8.3 Specific Criteria for Protection of Alternating Current Systems

Switchgear and Bus Protection

Consideration should be given to the higher fault current that may exist during parallel operation when selecting protective devices. If the power distribution system design cannot accommodate paralleling of bus supplies, or if paralleling can be allowed only under certain conditions, interlocks or procedural restrictions should be provided to restrict such paralleling.

If the power distribution design allows for automatic bus transfers, consideration should be given to the impact of the bus transfer on the coordination of protection devices. (See references [27] through [30].)

Bus Voltage Monitoring Schemes

Bus voltage monitoring schemes that are used for disconnecting the preferred power source, load shedding, and starting the standby power sources are part of the protection and should meet the criteria that follow:

a) Bus voltage should be detected directly from the Class 1E bus to which the standby power source is connected.

b) Upon sensing preferred power supply degradation, the condition should be alarmed in the main control room. On sensing preferred power supply degradation to an unacceptable low voltage condition, the affected preferred power supply should be automatically disconnected from the Class 1E buses.

c) Each division should have an independent scheme of detection of degraded voltage and loss of voltage conditions. Within each division, common equipment may be used for the detection of both conditions.

d) Each scheme should monitor all three phases. The protection system design should be such that a blown fuse in the voltage transformer circuit or other single phasing condition will not cause incorrect operation, nor prevent correct operation of the scheme. Means should be provided to detect and identify these failures.

e) The design should minimize unwanted operation of the standby power sources and disconnection of the preferred power supply. The use of coincident logic and time delay to override transient conditions is recommended.

f) Capability for test and calibration during power operation should be provided.

g) The selection of undervoltage and time delay set points should be determined from an analysis of the voltage requirements of the Class 1E loads at all on-site distribution levels.

h) Indication should be provided in the control room for any bypass incorporated in the design.

Protection of Motors and Feeder Circuits

The feeder circuit protection should consider any expected operating conditions of the motor that may require electrical system demands above the motor's nameplate rating such as the following:

a) motor service factor,
b) pump runout conditions,
c) operation at other than rated voltage.

Emergency Diesel Generator Protection

For diesel generator protection recommended practice, refer to IEEE Std 242 and ANSI C37-102 [21].

In the manual control mode, synchronizing interlocks should be provided to prevent incorrect synchronization whenever a standby power source is required to operate in parallel with the preferred power supply.

When a standby power supply is being operated in parallel with the preferred power supply, protection should be provided to separate the two supplies if either becomes degraded to an unacceptable level. This protection should not lockout or prevent the availability of the power supply that is not degraded.

The generator neutral grounding generally includes an arrangement of a resistor inserted in the neutral of the machine. The ground fault protection is normally provided by a relay that senses the passage of ground fault through the neutral grounding resistor. To limit high transient overvoltages due to ferroresonance, the resistance of the resistor should be low enough to limit the fault current to a few amperes. Under these low current conditions, the differential protection will not detect the fault and automatic tripping will not occur. Even though the fault will be very low, some welding of the generator stator laminations will occur. Also, in the presence of the fault, the voltage to ground of other parts of the stator windings will rise to $\sqrt{3}$ times normal. The voltage increase may cause a second ground fault to develop and a phase-to-phase fault might result, which could case very considerable immediate damage rendering the generator unusable and causing a fire which could damage all safety equipment in the affected emergency diesel generator (EDG) room.

Burning of generator iron is caused by the thermal effect of the ground fault current (I^2t), and while an accurate estimation is difficult, it could be theorized that burning of stator iron will start occurring after 2 min of initiation of the fault. After 10–15 min considerable burning may occur, and it could be estimated that restacking of the stator magnetic sheets may be required (a major repair). After 30 min, there would be a danger of a major fire event, which would become more possible by a likelihood of a second ground fault.

The relay utilized for the ground fault detection must be sensitive to fundamental voltage frequency and insensitive to third harmonic and multiples of third harmonic voltages.

The short circuit faults for the EDG can be calculated by using symmetrical components. For convenience, some of the formulae are given below, applicable to conditions of negligible resistance and unloaded generator. For further information and to understand the development of the formulae, please review symmetrical components theory in the references given at the end of this chapter.

Applying symmetrical components:

Generator three phase fault : If $= 1/X_1$,

where X_1 is positive sequence reactance.

Generator line-to-ground fault : If $= 3/(X_1 + X_2 + X_0)$,

where X_1 is positive sequence reactance, X_2 is negative sequence reactance, and X_0 is zero sequence reactance.

Generator line-to-line-to ground fault: If

$$3I_{a_0} = 3(V_{a_0}/X_0),$$

where V_{a0} is voltage of the zero sequence network,

$$V_{a1} = V_{a2} = V_{a0}$$
$$V_{a1} = 1 - I_{a1} * X_1, V_{a2} = -I_{a2}X_2, V_{a0} = -I_{a0}X_0$$
$$V_a = V_{a1} + V_{a2} + V_{a0}$$
$$V_b = a^2 V_{a1} + a V_{a2} + V_{a0}$$
$$V_c = a V_{a1} + a^2 V_{a2} + V_{a0}$$

Generator line-to-ground fault when there is a reactance inserted in the neutral connection to ground:

$$\text{Fault if} = 3/(X_1 + X_2 + X_0 + 3X),$$

where X is the reactance inserted in the neutral connection to ground.

Load Shedding and Sequential Loading

An automatic load shedding and sequential loading scheme should be included to ensure that the preferred or standby power sources can be loaded while maintaining voltage and frequency within acceptable limits.

The Class 1E bus load shedding scheme should automatically prevent shedding during sequencing of the emergency loads to the bus when connected to the standby power source.

If the preferred or standby power source breaker is tripped during or subsequent to loading, the load shedding and sequential loading scheme should be arranged to be automatically reset to perform its function in the event that the loads are to be reapplied.

Protection against Unbalanced Voltages

The principal effect of unbalanced voltages on motor operation is excessive temperatures in parts of the rotor, which result from excessive unbalanced (negative-sequence) currents. Guidance for derating induction and synchronous motors due to voltage unbalance is provided in National Electrical Manufacturers Association (NEMA) MG-1, Parts 20 and 21, respectively. Negative-sequence currents reduce the available accelerating torque. This lengthens the accelerating time and further contributes to motor overheating (IEEE PC37.96, Motor Protection, par 5.3).

Negative-Sequence Protection

Negative-sequence current is contributed by the motor or system when

- an unbalanced voltage condition exists (e.g., open-phase conditions, single-phase faults, or unbalanced load),

- stator coil cutout occurs during a repair, and
- there are shorted turns in the stator winding.

These negative-sequence currents induce double line-frequency currents that flow in the damper or rotor parts. The magnitude of the double line-frequency current depends on the location of the fault, number of turns shorted, mutual induction, and system and motor impedance. The danger to the rotor parts is a function of the unbalance in the stator current.

Unbalanced voltage and phase failures are similar phenomena, differing only in the degree of unbalance. While unbalanced phase voltages or currents are readily identified, it is the negative-sequence component that actually jeopardizes the motor. Hence, simple unbalance measurements may not provide the degree of motor protection required.

When the voltages supplied to an operating motor become unbalanced, the positive-sequence current remains substantially unchanged, and a negative-sequence current flows due to the unbalance. If, for example, the cause of the unbalance is an open circuit in any phase, a negative-sequence current flows that is equal and opposite to the previous load current in that phase. The combination of positive- and negative-sequence currents produces phase currents of approximately 1.7 times the previous load in each sound phase and zero current in the open phase. Because of additional motor losses, the actual value of the motor phase current in each sound phase is closer to twice the previous load current.

Three-phase voltages will still be observed at the motor terminals with one supply phase open. The actual magnitudes depend on the motor shaft load and on whether any other loads or capacitors are connected in parallel.

A small-voltage unbalance produces a large negative-sequence current flow in either a synchronous or induction motor. The per unit (pu) negative-sequence impedance of either type of motor is approximately equal to the reciprocal of the rated voltage pu locked-rotor current. When, for example, a motor has a locked-rotor current of six times rated, it has a negative-sequence impedance of approximately 0.167 pu on the motor-rated input kilovolt ampere base. When voltages having 0.05 pu negative-sequence component are applied to the motor, negative-sequence currents of 0.30 pu flow in the windings. Thus, a 5% voltage unbalance produces a stator negative-sequence current of 30% of full-load current. The severity of this condition is indicated by the fact that with this extra current, the motor may experience a 40–50% increase in temperature rise.

The increase in loss is largely in the rotor. Negative-sequence phase currents produce a flux that rotates in a direction opposite to the rotor rotation. This flux cuts the rotor bars at a very high speed and generates a pronounced voltage, resulting in a large rotor current. In addition, the 120-Hz nature of the induced current produces a marked skin effect in the rotor bars, greatly increasing rotor resistance. Rotor heating is substantial for minor voltage unbalance. Excessive heating may occur with phase current less than the rated current of the motor.

When a three-phase induction or synchronous motor is energized and one supply phase is open, the motor will not start. Under these conditions, it overheats rapidly and is destroyed unless corrective action is taken to deenergize it. The heating under these circumstances is similar to that in a three-phase failure to start, except that the line current is slightly lower (approximately 0.9 times the normal three-phase, locked-rotor current).

Unbalance protection must sense damaging conditions without responding to conditions for which the protective equipment is not intended to operate. Several classes of relays are used to provide unbalance protection.

Phase-balance relays (46) compare the relative magnitudes of the phase currents. When the magnitudes differ by a given amount, the relay operates. When an open circuit occurs on the load or source side of the CT supplying the relay, sufficient unbalance should exist to make it operate.

Electromechanical phase-balance relays can have a 1.0-A sensitivity and operate if one phase of the supply to the motor opens with the load on the motor if prior to the open the load current is in excess of approximately 0.6 A in the relay. Caution must be exercised in CT (Current Transformer) selection to ascertain with older phase- balance relays that the thermal capability of the relay is not exceeded at maximum load. Phase-balance relays without additional time delay beyond what the available relays inherently provide may cause unnecessary tripping of large motors during phase-to-ground or phase-to-phase disturbances remotely located on the power system. In as much as the clearing time of the relays on a power system is generally short in comparison to the required clearing time of phase-balance relays on the motor, a timing relay can be used without degrading the protection of the motor. Relay coordination is usually easier to attain with inverse-time relay characteristics than with separate fixed-time delays. Negative-sequence current relays (46) respond to the negative-sequence component of the phase currents. The instantaneous version of this relay provides excellent sensitivity.

Because of this, it will operate undesirably on the motor contribution to unbalanced faults on the supply system and, therefore, must trip through a timer or be directionally supervised.

Phase-balance relays need to incorporate time delay to ride through external unbalanced faults and other unbalanced disturbances.

The application of phase-balance and negative-sequence overcurrent relays [and neutral overcurrent relays (51 N)] should consider CT characteristics under high-current conditions. Excessive CT burdens result in current transformers saturating during fault conditions. Heavy motor-starting currents can also cause CT saturation, especially when the starting current has a large asymmetrical DC component. This DC resulting from motor inrush may last for a significant period of time, compared to an asymmetrical fault-current condition. This is because of the much greater inductive to resistance of the total circuit when starting a motor.

CT saturation is generally different in each phase because the asymmetry is different in each phase. This results in false negative- or zero-sequence currents. Unequal burdens in each phase can also cause unequal saturation between phases. Older phase unbalance relays usually have considerable variation in the burden between the three phases. This and any unsymmetrical conditions, such as single-phase ammeters and ammeter selector switches, can contribute to unequal saturation and possible relay misoperation.

CT saturation is minimized by keeping burdens low (especially the DC resistive component), by using high ratio current transformers, and by selecting current transformers with a high knee-point (saturation) voltage. Even these may be insufficient for the sustained DC in some motor inrush currents. In such cases, desensitizing the relay with respect to current, or increasing operating time, or both, may be necessary.

Negative-sequence voltage or reverse-phase relays respond to single phasing, to unbalanced voltage, or to reversed phase sequence. For motor protection, these relays must sense the same voltage supplying the motor. They are particularly applicable to a bus with substantial static load along with the motor load. For an all-motor load, the negative-sequence voltage relay may not, depending on the motor characteristics, operate for single-phasing at light load. Where motors constitute only a small proportion of the total load, single-phasing of the total load is recognized by this relay, even with no shaft load, irrespective of motor characteristics. In general, only the motor loads should be tripped when source single-phasing or excessive unbalance is recognized.

Negative-sequence voltage or reverse-phase relays respond to single phasing, to unbalanced voltage, or to reversed phase sequence. For motor protection, these relays must sense the same voltage supplying the motor. They are particularly applicable to a bus with substantial static load along with the motor load. For an all-motor load, the negative-sequence voltage relay may not, depending on the motor characteristics, operate for single-phasing at light load. Where motors constitute only a small proportion of the total load, single-phasing of the total load is recognized by this relay, even with no shaft load, irrespective of motor characteristics. In general, only the motor loads should be tripped when source single-phasing or excessive unbalance is recognized.

Negative-sequence Current Relay (Device 46)

These relays respond to the negative-sequence component of the phase currents and are available in the instantaneous type and the inverse-time type. When negative-sequence relays are used for the Device 46 function, an inverse time over current characteristic is usually used. These relays generally have an $I_2^2 t = k$ type of characteristic. That is, their time of operation is inversely proportional to the square of the negative-sequence component in the three-phase current. This type of relay inherently lends itself to proper coordination, even with many identical motors on one bus. Motor standards have not established values for k; however, a value of $k = 40$ has been used. A negative-sequence pickup setting equal to 15% of motor full-load current typically provides reasonable motor protection. This will trip at a 3% negative-sequence bus voltage for a motor with a typical 20% negative-sequence impedance. This threshold condition will result in an increase in motor losses approximately 10–25% of normal full-load losses of the motor (but are not related to motor load).

Even if the thermal protection provided takes into account negative sequence component of the current, it will not account for the additional heating due to high unbalance rate. In the event of the motor losing one phase of its supply, considerable overheating would occur, hence protection for negative sequence should be employed separately.

Recommended time delay of 10 s should provide adequate setting in most applications.

Device 47: Phase-Sequence or Phase-Balance Voltage Relay

Device 47 is a relay that functions upon a predetermined value of polyphase voltage in the desired phase sequence, or when the polyphase voltages are unbalanced, or when the negative phase-sequence voltage exceeds a given amount.

This device is similar in its function to that of Devices 27 and 59, the single-phase voltage relays. However, being a three-phase device, it responds to the three-phase quantities of the supply system. A relay responsive to the positive- or negative-sequence component of the applied voltage satisfies this definition and most of the needs in this area. However, electromechanical positive- or negative-sequence relays are sensitive to line frequency and, hence, the setting should make allowance for the specific relay in question. This is not a significant problem when the relay is used mainly to prevent attempting to start the motor with one phase missing or with reverse-phase sequence. A 90% setting is typical for a positive-sequence voltage relay. For a negative-sequence voltage relay, 5% is a common setting. However, it should not be assumed that any 47 device will prevent insulation deterioration during all possible unbalanced conditions.

A more common type of relay used for the Device 47 function is built on the principle of a three-phase induction motor. Such a relay has a torque proportional to the area within the voltage triangle. With balanced voltages, this is proportional to the positive-sequence voltage squared. As such, the relay is usually set to close its high-voltage contact to permit starting a motor at

90–95% of rated value. The undervoltage contacts are usually set to close at 80% of normal voltage. The control action that is initiated by the undervoltage contacts depends on the application.

When the three-phase voltages are not balanced, the area of the voltage triangle is no longer proportional to the positive-sequence voltage squared. The torque is now proportional to the difference between the positive-sequence and negative-sequence values squared. Thus, a condition with 90% positive sequence and 10% negative sequence would result in an effective voltage of 89% of normal. Usually, an operating time setting of 2 s upon complete loss of voltage is adequate for annunciation or to initiate the desired shutdown procedure.

8.4   Degraded Voltage Protection

(See IEEE 741 for additional information)

This section provides considerations for the following:

a) protection of Class 1E equipment for loss of voltage or degraded voltage conditions and
b) determination of the proper settings for loss of voltage and degraded voltage protection systems and their associated time delays.

The section also provides a description of a protection scheme utilizing solid-state undervoltage relays that will meet these requirements, including considerations for determination of the relay voltage set points and their associated time delays. However, it is recognized, because of the diversity of nuclear plant auxiliary system designs, that there are other protection schemes that will provide the desired level of protection. This section discusses the philosophy behind the desired actuation times and voltage levels. For each voltage and time delay relay, there should be applicable criteria to determine the set points along with the relay tolerance bandwidth.

Discussion

It is recommended that two levels of undervoltage detection and protection be provided on the Class 1E electrical distribution system. The first level of undervoltage protection is provided by the loss of voltage relays whose function is to detect and disconnect the Class 1E buses from the preferred power supply, the second level is provided by the degraded voltage relays that are set to detect a low-voltage condition. These relays alarm to alert the operators to the degraded condition and disconnect the Class 1E buses from the preferred power supply if the degraded voltage condition exists for a time interval that could prevent the Class 1E equipment from achieving its safety function or from sustaining damage due to prolonged operation at reduced voltage. The combination of the loss of voltage relaying system and degraded voltage relaying system provides protection for the Class 1E distribution system for conditions of

1. total voltage collapse or
2. sustained voltage degradation.

Determining the set point for degraded voltage relays usually requires a detailed analytical basis because the set point value is the result of a balance between preventing damage to Class 1E equipment and unavailability of Class 1E equipment. This must be achieved in a manner that avoids nuisance trips of the relay and resultant starts of the diesel generators due to an overly conservative set point.

Undervoltage conditions can be caused by equipment failure, concurrent motor starting, or bus transfers, in addition to low voltage on the preferred power supply. The set point calculations for degraded voltage relays should assure that the limiting equipment voltage requirements of the Class 1E system are met.

Degraded Voltage Relay Voltage Settings

The basis for the settings for the degraded and loss of voltage relaying systems should be supported by the following analyses that

a) calculate the minimum voltages at Class 1E buses for various operating and accident loading conditions,
b) determine the terminal voltages at the bounding components during worst case loading conditions so that they may be compared to minimum equipment voltage requirements, and
c) account for tolerances.

These analyses should also consider optimization of system voltage levels by appropriate adjustment of the fixed voltage tap settings of transformers, and for each applicable connection to the preferred power supply. Additionally, analyses should consider the effects of voltage compensating equipment, such as automatic load tap changing transformers and automatic switched capacitor banks, including their associated time delays, to ensure bus voltage recovery following expected voltage transients.

The capability to start motors, transfer buses, and ride through other momentary voltage dips must be evaluated. This evaluation should be performed at or below the lowest expected preferred power supply voltage. If analysis determines that the bus voltage drops into the degraded voltage relay operating range during a momentary voltage dip, the voltage must recover above the reset value within the time delay period.

Degraded Voltage Relay Time Delay Settings

After the voltage set point for the degraded voltage relays has been established, additional analysis is required to determine the appropriate time delays. These analyses should involve investigation of transient conditions, such as block motor starting and the effect of increased load currents from degraded voltage operation, on both protective device operation and equipment thermal damage. Two time delays should be determined.

a) *The first time delay* should be of a duration that establishes the existence of a sustained degraded voltage condition (i.e., longer than a motor starting transient). Following this delay, an alarm in the control room should alert the operator to the degraded condition. The subsequent occurrence of an accident signal should immediately separate the Class 1E distribution system from the preferred power supply.

In some cases, the voltage may be acceptable, but if an accident were to occur, the starting of large motors could cause the degraded voltage relays to begin timing. In this event, the time delay must be long enough to ride through the starting of accident loads without separating from the preferred power supply.

The time delay must not be so long that the system timing assumptions identified in the plant safety analyses are not met in the event there is a continued degradation of the preferred power supply.

b) *The second time delay* should be of a limited duration such that the permanently connected Class 1E loads will not be damaged or become unavailable due to protective device actuation. Following this delay, if the operator has failed to restore adequate voltages, the Class 1E distribution system should be disconnected from the preferred power supply. Protective devices (i.e., circuit breakers, control fuses, etc.) for connected Class 1E loads should be evaluated to ensure that spurious tripping will not occur during this time delay period. Consideration should also be given for restarting/reaccelerating the loads, should transfer to the alternate or standby power source be required

Loss of Voltage Relay Settings

The loss of voltage relays are an integral part of the undervoltage protection scheme, and their settings should be selected to limit the magnitude and duration of an undervoltage condition on the Class 1E buses. The function provided by these relays is to prevent equipment degradation by disconnecting the Class 1E buses from the preferred power supply in the event of collapsing or total loss of voltage. Voltage and time delay analyses should be performed to ensure that nuisance trips of the relays will not occur from anticipated dynamic effects such as motor starting or transmission system transients such as faults and switching.

High Voltage Conditions

Although degraded voltage is generally considered as an undervoltage condition, it also applies to a condition of unacceptably high voltage that can lead to equipment insulation damage or misoperation such as excessive motor shaft torque. The Class 1E system should be analyzed to determine the effects of high voltage. Rather than using design basis accident loadings, minimum unit loads such as those that may occur during cold shutdown or refueling modes may be appropriate in conjunction with maximum expected system voltage conditions.

In an overvoltage condition, an alarm is generally adequate, without automatic tripping, because such a condition would be expected to only cause gradual component loss of component life.

Off-Site System Voltage Considerations

It is important that acceptable transmission system voltages be determined so that, in the event of an accident, there will be assurance that the design basis loads can be supported without actuating the degraded voltage relays. Analyses should apply to each applicable off-site power supply configuration.

The required off-site voltage should consider transient variations in the off-site power system voltage levels caused by a trip of the nuclear unit and the possible bus transfers. Limits for high off-site power system voltages can be similarly determined by evaluating light load conditions. Transmission system studies should be performed in accordance with IEEE Std 765 [16].

Tolerances

Evaluation of set points should include the effects of tolerances that are affected by the following:

a) Operating tolerances
   1. ambient temperature variations;
   2. relay control power variations;
   3. accuracy class of potential transformers;
   4. repeatability of the relays.
b) Setting tolerances
   1. meter accuracy;
   2. meter calibration tolerance;
   3. setting tolerance permitted by procedures.

The operating tolerances are those elements that will cause the relay to operate at a value other than that to which it was set, while the setting tolerances are those that will cause it to be set at a value other than that which was intended.

A methodology for calculating tolerance effects is given in ANSI/ISA S67.04 [1]. The tolerance that is calculated should be added to the ideal set point to ensure that if the relays are operating on the low side of the tolerance band there will still be adequate terminal voltage for all Class 1E equipment. A margin may also be applied at this stage to achieve the desired set point.

8.5 Surge Protection

For surge protection of equipment and systems, refer to IEEE 141 and IEEE 242. For surge protection of induction motors, refer to IEEE C37.96. For guidance on protection of wire line facilities used for protective relaying and data transfer circuits, refer to IEEE 487 [13].

For recommendations in design and installation of low-energy, low-voltage signal circuits associated with solid-state electronic equipment, refer to IEEE 518 [14].

Surge protection should be provided to protect the shunt field of DC valve actuator motors. This surge protection may take the form of a resistor in the motor control center-wired in parallel with the shunt field to provide a discharge path for the shunt field's inductive voltage surges.

For guidance in the application of surge arresters to all types of power circuits and equipment, refer to IEEE C62.2 [22]. Refer to IEEE C62.41 [23] for guidance in determining the surge voltage for low voltage equipment, and to IEEE C62.45 [24] to provide guidance for tests that should be used to determine the surge withstand capability of the equipment used on low-voltage circuits.

8.6 Protection for Instrumentation and Control Power System

For guidance on protection for inverters, refer to IEEE 446 [12]. For information on ground protection practices, refer to IEEE 142 [3] and IEEE Std C62.92.3 [25]. For criteria for isolation and separation of non-Class 1E circuits from Class 1E circuits, refer to IEEE 384 [9].

Where a rectifier-type power supply is used as a source for an inverter, it should be provided with reverse current protection, current-limiting features or overload protection, and output undervoltage and overvoltage protection.

The instrumentation and control power distribution system should be provided with coordinated protection. Since inverters and motor generator sets are sources of limited short circuit current, special attention must be given to integrating protective device sensitivity and system available fault current. Coordination should include the protective devices in the alternate supply, inverters, static switches, distribution panels, instrumentation panels and racks, and other equipment powered from the system.

Where an instrumentation and control power bus is supplied by an inverter with current limiting characteristics and an automatic transfer has been provided to an alternate source with higher available current, this alternate source may be used to achieve the coordinated protection described above.

Ground detection monitoring to provide an automatic alarm should be included for ungrounded systems.

The instrumentation and control power system should be provided with undervoltage, overvoltage, and underfrequency protection. Where its power is supplied from a static inverter, overfrequency protection should also be provided. (For recommended practice on alarms and indication, refer to IEEE 944 [17].)

8.7 Protection Aspects for Auxiliary System Automatic Bus Transfer

This section gives a brief description of station auxiliary system automatic bus transfer schemes followed by a more detailed discussion of the impact to protection schemes and protective devices.

In this section, the primary emphasis is on bus transfers between two bus circuits under automatic control. These two circuits are designated in this section as normal and alternate sources, as identified in the figure.

Discussion

The subject of auxiliary bus transfer has been covered in many papers, and the concerns for motor torque and transformer impact loading are well documented. This section is intended to cover protection- related issues and specifically protection-related issues of nuclear generating plant station auxiliary systems. There are many concerns associated with bus transfer, some of which include the following:

a) damage to the motors being transferred (transient torque),
b) reacceleration after transfer of the motors being transferred,
c) ride through of any motors already supplied from the alternate source,
d) transfer coincident with starting of motors already supplied from the alternate source,
e) dynamic loading on the alternate source transformers,
f) alternate source overcurrent relay coordination,
g) saturation of differential circuit current transformers on the alternate source transformers,
h) underfrequency relay operation, and
i) undervoltage relay operation.

The basic protection schemes discussed in this section are shown in the one-line diagram of the transfer scheme.

An auxiliary system bus transfer is simply the transfer of power sources from the normal source to the alternate source. Many different schemes are in use, but most fall into one of the following categories and subcategories:

a) *Fast (4 –10 cycles) transfer* (the objective is to reduce bus dead time):
b) *Sequential.* This scheme uses *b* or early *b* contacts from opening of normal source breaker to close alternate source breaker.
   i) *Supervised.* This scheme uses a sync check type of relay to qualify transfer and is usually backed up by a residual scheme for cases where the sync check relay blocks fast transfer.
   ii) *Unsupervised.* The transfer is initiated without a sync check. This is a faster scheme but has exposures to out-of-phase switching conditions.
c) *Simultaneous.* This scheme uses the same control signal to trip the normal source breaker and close the alternate source breaker and relies on the fact that the breakers trip in less time than they close. This is a faster scheme, but there is no indication that the normal source breaker has opened before the alternate source closes.
d) *In-phase transfer* (the objective is to reduce shock to motors and system inrush by closing the alternate source bus breaker when the alternate source bus phase angle is within acceptable limits).
e) *Residual (40 + cycles) transfer* (the objective is to close alternate source breaker after the transferring bus residual voltage has decayed to a safe value).
f) *Voltage decay.* Alternate source breaker is closed when the transferring bus voltage has decayed to a predetermined value, usually around 25%.
g) *Time delay.* Alternate source breaker is closed based on a time delay that is associated with the expected voltage decay of the transferring bus.

The details of the different bus transfer schemes are very important to the operation of the systems associated with the motors involved in the transfer. However, the protection concerns are common to most of the automatic transfer methods and the following discussions are generic regardless of the transfer scheme.

Bus Transfer Analysis

Dynamic modeling is one method that can be used to predict the operation of the bus being transferred before, during, and after the transfer, to review protection concerns. Regardless of the modeling/analysis method used, many of the criteria in this section are applicable to the industry concerns.

a) *Pretransfer Case:* This case is used to establish the steady-state conditions of the motors prior to the transfer. At some point after a bus transfer signal is initiated, the source breaker is signaled to open. After the normal source breaker contacts begin to part an arc is formed. When the arc is extinguished the motor bus is operating as an isolated system.

b) *During Transfer Case* This case models the system as it operates isolated from a source. The time between extinction of the arc in the normal source breaker and reenergization from the alternate source breaker is commonly referred to as the "dead" time. During the dead time, the group of motors still connected to the bus will operate as an isolated system. Some of the motors in the group will act as generators, and others will act as loads. Eventually, the residual flux in all of the motors will be depleted and the bus voltage will drop to zero. In the interim, the composite bus voltage will begin to decay in both magnitude and phase. Accurate prediction of this decay is important in the design of the bus transfer scheme as well as the protection concerns.

c) *Post Transfer Case:* This case determines what dynamic voltages are present and currents flowing after the transfer. These voltages and currents are the primary concern for bus transfer protection issues.

The prediction of the voltage (magnitude and phase angle) across the alternate source breaker just prior to its closure (transition from the "during" to "post" cases described above) is referred to as preclosure voltage and is one of the three key elements in evaluation of the transfer scheme.

The four factors used to determine the voltage and current seen by the protective devices are as follows:

a) preclosure voltage (magnitude and phase angle),
b) alternate source voltage (magnitude and phase angle),
c) system impedance, and
d) motor bus voltage (magnitude and phase angle).

The transferring bus voltage is decaying in both voltage magnitude and phase angle. The voltage decay is primarily a function of the motor open circuit time constant. The frequency decay is primarily a function of the motor pretransfer loading and inertia. The heavier the motor is loaded prior to the transfer, the faster the decay during the transfer. The lower the inertia of the motor and load equipment, the faster the decay.

Analysis of a bus transfer scheme involves determination of voltages and currents after the transfer. Some factors to consider when performing these studies are as follows:

a) *Applicable to all types of bus transfer studies*
   1. Higher alternate source pretransfer bus voltage gives higher inrush current.
   2. Higher alternate source pretransfer bus voltage gives shorter motor reacceleration time.
   3. Motors should be modeled at actual load, or, if these data are unavailable, full load.

4. Motor and load H constants should be conservatively assumed.
5. If another bus is already supplied from the alternate source, simultaneous or delayed starting of a motor or motors on its bus should be reviewed coincident with the bus transfer if it is possible that other loads could be started for the mode being analyzed.
6. The decay of the motor bus voltage and frequency may result in operation of undervoltage or underfrequency relays on the bus or individual loads being transferred.
7. The impact of the power factor of the loads being transferred should be considered. For example, seasonal changes resulting in large amounts of resistance heating could impact the overall bus power factor and bus transfer study results.

b) *Applicable to fast bus transfer studies*
1. For fast transfer cases the maximum dead time should conservatively include the breaker's timing tolerance.
2. In fast transfer cases the phase angles of the normal source and alternate source bus voltages should be adjusted to achieve the worst case scenario when the alternate source breaker closes. This may involve consideration of the following:
   i) assumptions about what initiated the bus transfer, that is, system fault, unit out of step, etc.
   ii) a review of the transmission switchyard supply, that is, Did the cause of the unit trip result in splitting of the switchyard buses? If so, what is the impact to the alternate source bus voltage and phase angle?
   iii) H *constant* (machine inertia constant), which is a representation of the energy stored in the rotor when operating at synchronous speed. The units are in kilowatt-seconds per kilovolt-ampere rating of the machine. The H constant is based on inertia of the rotor plus load plus couplings. For the purpose of bus transfer, higher H constants result in slower voltage decay rates and lower H constants result in faster decay rates. Typical values for motors driving pumps are in the range of 0.4 and for large fans are 8.0.
3. Consider the loss of the worst case motor on the transferring bus during the transfer. For fast transfer cases tripping a high inertia load results in more rapid in frequency decay, which is the worst case.

c) *Applicable to residual bus transfer studies*
1. Consider the loss of the worst case motor on the transferring bus during the transfer. For residual transfer cases, tripping a low inertia load results in a slower decaying bus voltage and frequency that may be the worst case.
2. In residual transfer studies, the alternate source breaker should be closed when the alternate source supply voltage and the transferring motor bus voltage are 180 degrees out of phase to consider the worst case.

Bus Transfer Protection Issues

Assuming that the bus transfer schemes and protection system are properly designed and coordinated, the protection systems should not operate as a result of the transfer. However, consideration may be given to blocking automatic bus transfers for abnormal operating conditions such as bus differential relay operations and stuck breaker operations. When these abnormal conditions do not exist, one of the primary protection concerns is misoperation of protection systems resulting in unintentional loss of the alternate source to either the bus being transferred, a bus already being supplied by the alternate source, or both. In many nuclear plants, the bus or buses already supplied by the alternate source are Class 1E. Therefore, loss of power to the bus as a result of a bus transfer is undesirable.

Current Transformer Saturation

The common concern associated with bus transfer schemes and current transformer saturation is the misoperation of transformer differential relays. When current transformers are selected, consideration should be given to the maximum primary fault current available for the anticipated grid and auxiliary system conditions. The performance of the selected current transformer should then be confirmed for the maximum asymmetric fault current by means of current transformer performance application guides or calculation. The concern in the transformer differential relay case is that the high-side current transformers saturate and the low-side current transformers do not saturate. The differential relay senses a difference in high- and low-side VA flow and operates to clear an apparent fault in the transformer (see IEEE C37.91 [19]). The result is loss of the alternate source.

The impact of possible saturation of current transformers due to high inrush currents associated with the closing of the alternate source breaker should be evaluated. A similar phenomenon is discussed in [29]. The basic problem with current transformer saturation is that the current transformer secondary current may no longer be a true proportional representation of the primary current. The inrush currents associated with bus transfer may be higher than seen for individual motor starting. Therefore, the operation of the current transformers and connected relays would need to be reviewed.

Overcurrent Relay Coordination

The concern in the overcurrent relay case is that the phase overcurrent relays on the alternate source feeder (the alternate source breaker is shown in Figure 8.1) might operate as a result of the bus transfer. If an instantaneous overcurrent element were in service, it might pick up on the inrush current associated with the transfer. In this case the setting of the instantaneous element

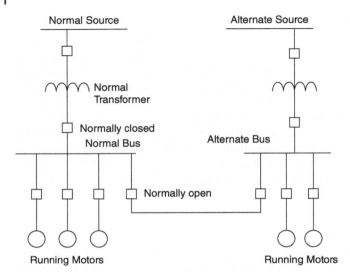

Figure 8.1 One-line diagram for bus transfer scheme.

must be set high enough that it does not operate for the worst case bus transfer inrush. The instantaneous setting of this relay is often based on a multiple of the largest single motor's locked-rotor current and coordinated with the overcurrent protection of the feeders to individual loads fed from the bus being transferred. The bus transfer inrush may exceed this setting value. If these relays were not properly set, operation of the instantaneous device could result in tripping of the alternate source breaker for an otherwise acceptable bus transfer.

Undervoltage and Underfrequency Relays
An additional protection concern involves undervoltage relays. As mentioned earlier, the alternate source for the bus being transferred may already be supplying other loads. In the case of nuclear plants, this may be a Class 1E bus. If the Class 1E bus were equipped with undervoltage relays, they would sense the voltage dip that occurs when the alternate source breaker closes. The recovery of the bus voltage needs to occur at a rate that does not result in actuation of the undervoltage relays. Otherwise, the result might be separation of this Class 1E bus from the power source which is not desirable.

The undervoltage and underfrequency relays may also appear on the auxiliary bus being transferred. In this case, their operation could impact the successful transfer of certain loads. If critical loads were being transferred that were desirable to be available immediately after the transfer, the unintended operation of these relays may defeat the advantages of the automatic bus transfer.

Implementation of a Bus Transfer Scheme Using Dynamic Modeling
a) Build a computer model of the involved system:
   1. develop distribution system impedance diagram;
   2. model motors at actual load, or, if these data are unavailable, at full load; and
   3. typically, motors are predominately squirrel cage induction type.
b) Simulate bus transfer scenarios using computer models:
   1. highest alternate source voltage gives highest inrush and V/Hz;
   2. lowest alternate source voltage gives longest reacceleration time;
   3. assume any single load on the transferring bus is tripped;
   4. determine bus dead time based on normal and alternate source breakers operating times including worst case tolerances;
   5. assume bus voltage is 180 degrees out-of-phase with an alternate source for residual transfer; and
   6. utilize the dynamic voltage, current, and frequency values determined from the analysis to review settings of protective devices.
c) Implement a recommended scheme.
d) Perform field testing of actual bus transfer scheme:
   1. test bus voltage decay first;
   2. verify relay operation and functional logic;
   3. validate results from computer model;
   4. determine any unintentional loss of loads during transfer.
e) Reverify or fine-tune analytical model:
   1. adjust the model parameters to match the test conditions as closely as possible;
   2. verify system voltage, phase angles, motor loading, breaker timing, etc.

Historically, the acceptability criterion related to the motors being transferred was known as the 1.33-V/Hz criterion. This first appeared in ANSI C50.41 and later in NEMA MG 1. The 1.33-V/Hz criterion was related to the vector difference in the transferring motor bus residual voltage and the alternate source bus voltage at the instant of transfer. More recent papers have shown that the 1.33-V/Hz criteria are not always reliable prediction of motor shaft torque (potential motor damage). As a result, NEMA MG 1 has been revised to remove the 1.33-V/Hz criterion (ANSI C 50.41 has been withdrawn). At the present time, there is not a generic acceptability criterion. There are several papers on proposed methods and research projects to determine a new acceptability criterion. Today utilities are faced with determining their own bus transfer acceptability limits.

Bus Transfers and High-Speed Magnetic Circuit Breakers
For the purpose of this chapter, high-speed breakers are defined as those capable of sensing and clearing faults within one cycle or less.

This section discusses the trip setting requirements of adjustable high-speed magnetic-only circuit breakers, also known as motor circuit protectors, to preclude inadvertent tripping during bus transfer, and motor jogging and plugging operations.

The time–current curves of the magnetic-only and thermal-magnetic types of circuit breakers are compared to provide a better understanding of their performance characteristics for motor feeder circuit fault protection.

Discussion

Short circuit protection devices in alternating current motor circuits consist of either thermal-magnetic or magnetic-only circuit breakers. In combination motor starters, the magnetic circuit breaker and the thermal overload relay (TOR) share a common enclosure. The magnetic trip setting of adjustable high speed magnetic-only circuit breakers, also known as motor circuit protectors, requires special considerations. The following discussion assumes that all other circuit breaker parameters, such as trip and interrupting ratings, are correctly chosen.

Ordinarily, spurious trips resulting from circuit switching transients are avoided by selecting an instantaneous trip setting equal to approximately two times the motor nominal locked-rotor current. This conventional basis for trip setting adequately accounts for effects such as manufacturing tolerances (20%), system asymmetry (50%), and overvoltage (10%).

However, a setting in this range can produce inadvertent tripping during bus transfer, jogging, and plugging operations resulting from the additional applied voltages that these operations may produce. Depending on the application, to avoid spurious tripping, the instantaneous trip setting on high-speed magnetic-only breakers may need to be set as high as twice the conventional value, that is, four to five times the motor-locked-rotor current. However, the setting should be reviewed to ensure that the thermal damage curves of the protected equipment (e.g., cable, penetration, motor) has not been compromised and that the setting is lower than the available short circuit current at the load.

For breakers with a low trip rating (e.g., 3 and 7 A), the control circuit inrush current can be a significant portion of the breaker setting. The setting should be high enough to ensure that spurious tripping does not occur.

When selected in this manner, settings may exceed conventional guidelines for instantaneous trip settings (maximum 1300% of motor full load current) in favor of completing the equipment safety function. However, it can be shown that with low trip rating magnetic-only breakers (3, 7, and 15 A), an instantaneous trip setting selected on the basis of four to five times motor nominal locked-rotor current provides greater motor circuit fault protection than possible with nonadjustable thermal-magnetic circuit breakers.

The magnetic trip region of a typical 15-A nonadjustable thermal- magnetic circuit breaker is 12–50 times the breaker rating, which is the smallest standard size, magnetic actuation begins at currents as low as 180 A or as high as 750 A. This indicates that the use of a 15-A thermal-magnetic circuit breaker in a motor circuit with motor full-load current less than 14 A may not meet conventional magnetic trip setting criterion.

8.8 Protection for Primary Containment Electrical Penetration Assemblies

An electrical penetration assembly should be considered as part of the cable system between the load and the primary interrupting device. For guidance in the application of electrical circuit protection, refer to IEEE 242 [4], which includes information also applicable to electrical penetrations. Short circuit, overload, and continuous current ratings and capabilities of the electrical penetration are defined in IEEE 317 [6].

The electrical penetration assemblies installed as part of the containment structure require special consideration in the selection of their protection. This special consideration arises where the potential exists for a fault inside containment to result in a penetration seal failure, such that a breach of containment may occur.

Electrical penetrations requiring special consideration (i.e., where protection is required to ensure containment integrity) should be provided with dual primary protection operating separate interrupting devices, or primary and backup protection operating separate interrupting devices.

The time–current curves of the dual primary protection or the primary and backup protection should coordinate with the time–current capability curve of the electrical penetration to be protected.

Designs that use dual protection for penetration circuits provide adequate assurance that the protection will function acceptably under all plant conditions. Since the plant design basis is that a seismic event will not cause a loss-of-coolant accident or high energy line break inside containment, seismic qualification of protective devices for non-Class 1E circuits that penetrate the containment is not required.

8.9 Protection of Valve Actuator Motors (Direct Gear Driven)

(Refer to IEEE 741 [15] for additional information and review references [31] through [33]).

This section analyzes the selection and setting of electrical protective devices for value actuator motors (VAMs); criteria for setting mechanical devices within the valve actuator, such as torque and limit switches, are not within the scope of this document. (For VAM applicable definitions, refer to Appendix 2).

Operationally, a valve is a device that has two end limits of travel: full open and full close. Typically, a valve actuator will have built-in position and/or torque switches to stop the unit at its full open or full closed positions.

While the torque switches may provide some degree of overload protection, they are only intended to ensure proper seating of the valve at its extreme positions. As such, there is always a need to provide overload protection to complement the torque switches. The torque switches will not protect the valve assembly, including the motor, for torques less than the setting of the torque switch. Furthermore, if the torque switches are set higher than the torque available from the motor, the motor will go into a stall or locked-rotor condition when the actuator movement becomes mechanically limited at the endpoints of the valve stem movement (full open or closed). Locked-rotor results in overheating of either the stator or rotor, depending on motor design. If the stator reaches its insulation temperature limit before the rotor reaches its limiting temperature, the motor is considered "stator limited," whereas, if the rotor reaches its maximum limit before the stator, the motor is considered "rotor limited." In midtravel, valve actuator motors may encounter torques substantially greater than the rated nominal torque due to mechanical conditions because of problems such as a nonlubricated stem or binding in either the valve disk/wedge or stem, which results in a running current that is higher than the rated nominal current.

The selection and the setting of the electric protective devices for the direct geared valve actuator motor should ensure the following:

- *Coordination with motor allowable temperature.* The time current characteristic of the protective device is coordinated with the time current characteristic of the motor, as derived from motor time temperature data.
- *Coordination with valve duty cycle.* The coordination should ensure the allowable duty cycle of the valve is completed without compromising the motor thermal withstand capability, while allowing margin for variations in current drawn by the motor, or in the thermal characteristics of the protective device, or both.
- *Prevention of motor overheating due to locked-rotor conditions.* To protect the motor during locked-rotor conditions, the protective device maximum trip time should not exceed the allowable safe locked-rotor time, and the minimum trip time should not be less than the acceleration time (typically less than 1 s).
- *Prevention of motor overheating due to anticipated overloads.* Protective devices should be coordinated with the motor allowable operating time corresponding to nominal torque and anticipated overloads. Typical anticipated

valve overloads fall in the range of 150–300% of the valve actuator motor nominal torque, depending on the actuator type and application.

- *Prevention of nuisance trips* during acceleration, due to anticipated overloads, and during operation within the duty cycle of the valve.
- *Short circuit protection* for the valve actuator motor. If the device for short circuit protection includes an overload element, this element should be coordinated with the valve actuator motor thermal overload device.

In calculating set points, the following should be considered:

- accuracy of the protective device (tolerance),
- effect of ambient temperature variation, and
- effect of motor terminal operating voltage extremes.

VAM currents necessary for the setting of the protective device are

- at nominal torque,
- at selected overload torque (150–300% of nominal torque), and
- at locked-rotor torque.

The current values should be obtained from the manufacturer or measured at nominal voltage and either measured or calculated for anticipated voltages at the terminals of the motor.

Overload Protection Devices

Guidelines associated with motor stator winding protection and overload protection devices are given in IEEE C37.96 [20] and are applicable for VAMs. The overload protection for VAMs is commonly provided by a combination of the following:

- *internal devices* located on stator windings,
- *external devices* actuated by motor current, and
- *combination* of internal and external devices.

Internal Devices

These sensors may be vulnerable to vibration and are not easily accessible for maintenance. Their use in power generating stations is usually limited to applications for alarm/surveillance purposes only. Internal temperature sensors provide adequate early warning for motors that are stator temperature limited, but are ineffective for rotor-limited motors.

Current Sensing Devices

External devices that are actuated by motor current include bimetallic TOR provided as part of motor-starters and solid-state overload (SSO) devices. The overload relays provide protection for overload and locked-rotor conditions.

The SSO devices may be microprocessor based or solid-state type with electronic circuitry. The SSO relays may be separately mounted on the motor starter/contactor or be an integral part of solid-state starters. The thermal model in these devices is based on the line current drawn by the motor and sense-through the built-in current transformers. The logic circuitry uses this current signal to calculate the protective characteristics. Several other optional protective features can be added to the overload protective function.

These devices are highly accurate—some of them have an accuracy of up to 1%. They are available in continuously adjustable NEMA classifications 5–30. This feature is extremely useful in selecting overload protection for VAMs with long stroke times due to compliance to the Nuclear Regulatory Commission Generic Letter 89-10 program or due to gearing changes on the VAM operator. With a built-in thermal memory function, these relays are capable of providing overload protection for a cold as well as a hot start of a motor. For the VAMs used in a cycling, jogging, or plugging application this is a particularly useful function. By selecting a relay with appropriate cold and hot start time–current characteristic curves, a motor can be protected during repeated starts.

IEEE Std C37.96 provides guidance for selection of TORs for small induction motors.

TORs made with bimetallic elements will have a tolerance band. The minimum tripping time should be considered to avoid a spurious trip during stroking of the valve, and the maximum tripping time should be considered to avoid exceeding thermal capability of the VAM.

Special Considerations for Valve Actuator Motor

NEMA ICS 2 provides information on overload heater time–current characteristics. Application of the NEMA standard to VAMs, however, requires special considerations. The most common practice is to rate ac VAMs for 15 min and DC VAMs for 5 min.

The current drawn by the VAM is variable in magnitude, and when the current is averaged over its entire stroke, its magnitude may be greater than the VAM nominal current. Therefore, based on industry experience, it is recommended that the current at twice nominal torque be used as a checkpoint for overload heaters to preclude tripping during the stroke. In addition, the time–current characteristics of the TOR should be coordinated with the time–current characteristics of the motor at nominal torque for duty cycle time, twice nominal torque for stroke time, and at locked-rotor. High-velocity butterfly valves may require higher average torque, and the TOR should also be checked at a current that corresponds to 2.5 times VAM nominal torque for one stroke to assure that no spurious tripping will occur.

VAM acceleration times are normally very short. As a result, spurious tripping during acceleration can normally be prevented if the minimum protective device trip time at the locked-rotor current exceeds 1 s. For certain VAMs, it may be difficult to meet the constraints of locked-rotor time and still meet the stroke or duty cycle time at anticipated overloads. Overload relay characteristics of different NEMA classifications and/or models should be evaluated to obtain optimal coordination. In the case of high-torque motors with special alloy rotors, there may be no margin in the safe locked-rotor times. DC VAMs have considerably shorter duty rating than ac VAMs. They require the use of faster acting TORs that will ensure satisfactory completion of the actuator's duty cycle at anticipated operating currents corresponding to actual torques encountered during the duty cycle period or twice the nominal torque as the case may be.

Determination of duty cycle should be based on system design requirements. To avoid spurious TOR actuation during the duty cycle of the VAM, nominal current should be checked for the duty cycle time for the selection of the TOR. Thermal heat buildup of the VAM by stroking in excess of the duty cycle may not be protected by the TOR.

Typical Motor Operated Valve Circuit Breaker Protection Options

Many valve actuator motors in nuclear generating plants have nameplate full-load currents below 12 A. It should be analyzed if breakers should be magnetic-only or thermal magnetic to optimize the protection.

Issues to be addressed include:

a) Greater motor circuit fault protection than possible with currently available thermal-magnetic circuit breakers. Note that a 7-A magnetic-only breaker, set at 70 A will trip at a maximum of 100 A, whereas a nonadjustable 15 A thermal-magnetic breaker may not trip in its magnetic region until 750 A.
b) Interrupting low magnitude short circuit faults by circuit breaker action provides greater protection against equipment damage in general, and the protection of TORs against misapplication as fault interrupters in particular.

Information Needed for the Selection and Coordination of Overload Relay Protection

The following information is required to select overload heaters:

a) VAM currents at rated voltage and expected minimum and maximum voltages:

  1. rated nominal current,
  2. current at twice the rated nominal torque or at a selected torque for which the corresponding thermal capability is available, and
  3. locked-rotor current.
       Note: Manufacturer or field list data, or both, for LRC may vary from the nameplate current.
  b) time temperature characteristics of motor:
  1. rated nominal current,
  2. time the motor can safely carry current corresponding to twice nominal torque, and
  3. locked-rotor duration the motor can safely tolerate.
  c) TOR time–current curves and TOR type and selection table and TOR application guidelines from the selected manufacturer or from IEEE C37.96;
  d) motor-rated ambient temperature, insulation class, rated nominal torque, and rated speed;
  e) stroke time of the valve actuator;
  f) maximum allowable duty cycle; and
  g) TOR ambient temperature during normal and abnormal plant operating conditions (if ambient compensated relays are not used).

Correction for Ambient Temperature

TORs are available either as ambient compensated or uncompensated. It is preferable to use ambient-compensated TORs, as these will essentially have the same minimum operating current and time–current characteristics within the range of its ambient compensation. For application of uncompensated TORs, specific manufacturers' recommendations should be obtained regarding time–temperature characteristics for additional guidelines.

Correction for Voltage Variation

Alternating current VAMs are sensitive to voltage variations because the magnetizing current of these motors represents a significant portion of the nominal rated current. As voltage increases, current increases; as voltage decreases, current decreases.

Terminal voltage variations affect DC VAMs as follows:

  a) *Locked-rotor current.* The magnitude of the locked-rotor current varies in direct proportion to the terminal voltage. The effect of higher current at higher voltage should be considered in the selection of the TOR in consultation with the manufacturer.
  b) *Current drawn during valve stroking.* Voltage variations have an insignificant effect on the current drawn by DC VAM during stroking. However,

they have an effect on the stroke time due to change in speed resulting in either a shorter stroke time for higher voltage, or a longer stroke time for lower voltage. As the variation in current drawn is negligible, higher voltages do not compromise the thermal time–current capability of the VAM and, as such, have no effect on selection of the TOR. However, lower voltage may result in longer operating time and, thus, have the potential to result in spurious operation of the TOR if not properly selected.

Typical Procedures for Selection of TORs

The procedures applicable to VAMs is discussed here are adapted from IEEE C37.96 to suit the unique requirements of VAMs.

It is desirable to set TORs to allow valve motors to carry overloads that will not damage the motor, since torque requirements are variable and stroke times are relatively short compared to rated operating times. For this reason, TOR operation should be selected to allow a motor to run at twice nominal torque during its stroke time or at nominal torque for the duty cycle time. It should be noted that actual TOR operation can be as high as 125% of the minimum current, or 115% of the maximum current of the indicated range of the TOR. To select a TOR, a multiplier (which recognizes TORs are designed for continuous applications, whereas valves are intermittent loads) should be used to reduce the value of the motor full load amperes. This information should be obtained from the TOR manufacturer.

The current corresponding to twice nominal torque and motor locked-rotor current is divided by the current corresponding to 1 pu value to arrive at pu currents, which will determine the trip time at overload and locked-rotor. If trip time is too short, the next larger heater should be evaluated.

After the trip time under locked-rotor condition is found adequate, the trip time for nominal torque and twice nominal torque is determined. An evaluation is made based on these trip times relative to the duty cycle of the valve and the thermal capability of the motor, as defined by the motor performance curve.

The above concepts, including the application of the criteria, are illustrated in the following procedure for thermal overload trip criteria:

a) When carrying locked-rotor current, the TOR should actuate in a time within the motor's limiting time for carrying locked-rotor current (typically 10–15 s).
b) When carrying twice nominal torque current, the TOR should actuate in a time within the motor's limiting time for carrying twice nominal torque current. (This time is to be obtained from the motor's temperature–time–load characteristic curve.)
c) When carrying the current at twice nominal torque, the TOR, as a minimum, should not actuate within the stroke time of the VAM.

d) When carrying nominal current, the TOR should not actuate within the duty cycle time of the VAM.
e) When carrying locked-rotor current, the TOR should not actuate in the first second to allow the VAM to accelerate.

8.10 Protection for DC Systems

For recommended practice on protection for batteries, refer to IEEE 946 [18]

The DC power distribution system should be provided with coordinated protection. Coordination for DC power system circuits should include the main bus protective devices and the protective devices used in branch circuits, in switchgear control circuits, and in relay and process control panels. Care should be taken to use appropriate correction factors or DC trip characteristic curves for protection devices.

For criteria for isolation and separation of non-Class 1E circuits from Class 1E circuits, refer to IEEE 384.

Ground detection monitoring should be provided for ungrounded systems.

Battery chargers should be provided with current limiting features or overload protection, reverse current protection, output undervoltage, and overvoltage alarms and/or trips. For additional guidance on the protection of battery chargers, refer to IEEE 446.

8.11 Testing and Surveillance of Protective Systems

Testing of voltage transformers, current transformers, relays, fuses, trip elements, and other devices should be in accordance with IEEE 141 [2], IEEE 336 [7], and IEEE 338 [8]. Where overload heater elements are used, they should be tested in accordance with NEMA ICS 2 [26].

Preoperational Tests

A program should be established to assure that preoperational testing will demonstrate the satisfactory performance of the protection, refer to IEEE 415 [11].

Requirements for initial preoperational testing will also apply after major modification or repair to the protection. The program should include checks, verification, tests, and reports to demonstrate that the following are true:

a) There is proper operation according to system design.
b) The protection will meet requirements relating to voltage, frequency, current, power, and other limits.
c) That where redundant power or control systems are installed, the failure or loss of one system will not prevent correct operation of the redundant system.

d) The failure of a non-Class 1E power or control system will not adversely affect the correct functioning of the protection for Class 1E equipment.
e) The specified requirements for the operating environment are not violated. These requirements may include cleanliness, temperature, humidity, and vibration.

Surveillance

For surveillance methods to demonstrate equipment operational status, refer to IEEE 308. For periodic test requirements pertaining to those parts of the protection that perform a safety function, refer to IEEE 338.

The protection should be designed to permit periodic testing to provide assurance that the protection can perform its function. This standard does not address the periodic testing of actuated electrical equipment.

Periodic tests should duplicate as closely as practicable the performance requirements of the actuation devices. Acceptable methods for periodic testing include the following:

a) Testing of each protection circuit from sensor through actuated equipment.
b) Testing part of each protection circuit and actuation device individually or in previously selected groups. In this instance, overlap requirements should establish an acceptable basis for combining individual or group test results.
c) Testing each electrical actuation circuit individually if the actuated equipment has more than one actuation device.

During periodic protection device testing, where the ability of the safety system to respond to an accident signal has been made inoperative, the following should apply:

a) The inoperative condition should be indicated in the main control room.
b) Means should be provided to prevent any unexpected operation of the bypassed device

Questions and Problems

8.1 A nuclear power plant has a loss of coolant accident, and it is mitigating the event from its connection to the preferred power supply. Immediately after the initiation of the event, a loss of off-site power ensues and the two redundant EDGs automatically start and load with the safety loads. Ten minutes into the event, a ground fault alarm is received for EDG-1, automatic tripping on ground faults was bypassed by design and the operator needs to decide on any manual actions required. (Note: solution of this problem may be based on the information provided with the notes; however, students are encouraged to expand with any available information from the bibliography in terms of level of damage caused by generator ground faults.)

Analyze the operator's action providing justification to your answer for each one of the cases below:

Case 1. The operator immediately shuts down EDG-1. Assume that the operator action took 90 s

a) Was the operator's action appropriate in terms of minimizing generator damage? Was damage to the generator stator iron most likely prevented?

b) Was the operator's action appropriate in terms of safeguarding the plant and minimizing the danger or radiation release?

Case 2. Because the operator was extremely busy with the event mitigation and the investigation of the generator fault, he was unable to shut down EDG-1 for 20 min.

a) Was the operator's action appropriate in terms of minimizing generator damage? Was damage to the generator stator iron most likely prevented?

b) Was the operator's action appropriate in terms of safeguarding the plant and minimizing the danger or radiation release?

c) What should have been the optimal response from the operator?

8.2 A nuclear power plant EDG has the neutral grounded through a resistor to limit the magnitude of the maximum ground fault. The resistor has a 10-min rating, which was chosen by the plant for economic reasons (the lower the time rating the less costly is the resistor).

A question was raised regarding the adequacy of the resistor time rating. Provide your opinion with justifications to your answer.

8.3 A nuclear power plant EDG has the neutral connected directly to earth.

a) As the direct neutral grounding to earth does not include the separate devices for increasing neutral connection impedance, would this generator have a higher reliability compared to one where the neutral is grounded through an impedance?

b) What will be the cost of this generator compared to one where the neutral is grounded through an impedance?

c) Most faults are phase-to-ground faults; therefore, on a probability basis will the generator be more prone to high damage type failures compared to one where the neutral is grounded through an impedance?

8.4 A nuclear plant EDG is rated 4250 kW, 0.85 power factor, 4.16 kV, and has a direct-axis subtransient reactance of 0.52 Ω. The negative sequence reactance is 0.3 Ω, and the zero sequence reactance is 0.25 Ω. The neutral of the generator is solidily grounded. For "a" phase-to-ground fault, when the generator is operating unloaded at rated voltage, neglecting resistance determine:

a) The subtransient line-to-ground fault current in the generator, in amperes

b) The line to neutral subtransient voltages, in volts Refer to IEEE 399 [10]

8.5 Calculate the magnitude of the fault in Amperes for line-to-line-to-ground for the generator of problem 8.4

8.6

a) Calculate the three phase fault in amperes for the generator of problem 8-4

b) Calculate the ratio of the three-phase fault to the fault calculated in problem 8-4

c) What can you comment on the calculation under (b) above?

8.7 For the generator of problem 8.4, calculate the ohmic value of a reactance to be inserted between the generator neutral and ground so as to limit the ground fault to the same value as the three-phase fault.

8.8 Following the guidance of IEEE 741, Annex B, select an overload relay heater for an ac VAM, for TOR manufacturer A and considering the following data:

Motor to be protected:

Starting torque	12 ft-lb
Speed	3400 rpm
Frequency, voltage, phases	60 Hz, 460 V, 1.4 HP
Nominal torque	3 ft-lb
Duty class	15 min, class B, totally enclosed, nonventilated effect of voltage: to be neglected for this problem
Nameplate nominal current	2.8 A
Current at twice nominal torque	3.6 A
Locked rotor current	18.0 A
Acceleration time	1 s
Valve stroke time	200 s
Duty cycle	Two strokes
Duty cycle time of valve actuator	360 s

(continued)

Protective device data	manufacturer A heater
Ambient compensated	Figure B.2 of IEEE 741
Heater selection table of TOR manufacturer A	Table B.1 of IEEE 741

8.9 Following the guidance of IEEE 741, determine the settings of the degraded voltage relay. Assume condition stipulated in IEEE 741.

References

1 ANSI/ISA S67.04 Part 1, "Setpoints for Nuclear Safety-Related Instrumentation."

2 IEEE Std 141, "IEEE Recommended Practice for Electric Power Distribution for Industrial Plants (ANSI)."

3 IEEE Std 142, "IEEE Recommended Practice for Grounding of Industrial and Commercial Power Systems (ANSI)."

4 IEEE Std 242, "IEEE Recommended Practice for Protection and Coordination of Industrial and Commercial Power Systems (ANSI)."

5 IEEE Std 308, "IEEE Standard Criteria for Class 1E Power Systems for Nuclear Power Generating Stations (ANSI)."

6 IEEE Std 317, "IEEE Standard for Electric Penetration Assemblies in Containment Structures for Nuclear Power Generating Stations (ANSI)."

7 IEEE Std 336, "IEEE Standard Installation, Inspection, and Testing Requirements for Power, Instrumentation, and Control Equipment at Nuclear Facilities (ANSI)."

8 IEEE Std 338, "IEEE Standard Criteria for the Periodic Surveillance Testing of Nuclear Power Generating Station Safety Systems (ANSI)."

9 IEEE Std 384-1992, "IEEE Standard Criteria for Independence of Class 1E Equipment and Circuits (ANSI)."

10 IEEE 399-1990, "Recommended Practice for Power System Analysis."

11 IEEE Std 415, "IEEE Guide for Planning of Preoperational Testing Programs for Class 1E Power Systems for Nuclear Power Generating Stations (ANSI)."

12 IEEE Std 446, "IEEE Recommended Practice for Emergency and Standby Power Systems for Industrial and Commercial Applications (ANSI)."

13 IEEE Std 487, "IEEE Recommended Practice for the Protection of Wire-Line Communication Facilities Serving Electric Power Stations (ANSI)."

14 IEEE Std 518, "IEEE Guide for the Installation of Electrical Equipment to Minimize Electrical Noise Inputs to Controllers from External Sources (ANSI)."

15 IEEE Std 741, "IEEE Standard Criteria for the Protection of Class 1E Power Systems."

16 IEEE Std 765, "IEEE Standard for Preferred Power Supply (PPS) for Nuclear Power Generating Stations (ANSI)."

17 IEEE Std 944, "IEEE Recommended Practice for the Application and Testing of Uninterruptible Power Supplies for Power Generating Stations (ANSI)."

18 IEEE Std 946, "IEEE Recommended Practice for the Design of DC Auxiliary Power Systems for Generating Stations (ANSI)."

19 IEEE Std C37.91, "IEEE Guide for Protective Relay Applications to Power Transformers (ANSI)."

20 IEEE Std C37.96, "IEEE Guide for AC Motor Protection."

21 IEEE C37-102, "Guide for AC Generator Protection."

22 IEEE Std C62.2, "IEEE Guide for Application of Gapped Silicon-Carbide Surge Arresters for Alternating Current Systems (ANSI)."

23 IEEE Std C62.41, "IEEE Recommended Practice for Surge Voltages in Low-Voltage AC Power Circuits (ANSI)."

24 IEEE Std C62.45, "IEEE Guide on Surge Testing for Equipment Connected to Low-Voltage AC Power Circuits (ANSI)."

25 IEEE Std C62.92.3, "IEEE Guide for the Application of Neutral Grounding in Electrical Utility Systems, Part III—Generator Auxiliary Systems (ANSI)."

26 NEMA ICS 2, "Industrial Control and Systems: Controllers, Contactors, and Overload Relays, Rated Not More Than 2000 Volts AC or 750 Volts DC."

Bus Transfer Schemes

27 R. H. Daughterty, "Analysis of Transient Electrical Torques and Shaft Torques in Induction Motors as a Result of Power Supply Disturbances," *IEEE Transactions on Power Apparatus and Systems*, vol. PAS-101, no. 8, pp. 2826–2834 (see discussion by Cummings), August 1982.

28 IEEE Power System Relay Committee, "Relay Performance Considerations with Low Ratio CTs and High Fault Currents," *IEEE Transactions on Power Delivery*, vol. 8, no. 3, pp. 884–897, July 1993.

29 J. S. C. Htsui, "Magnitude, Amplitude and Frequencies of Induction-Motor Air Gap Transient Torques through Simultaneous Reclosing without Capacitors," *IEEE Transactions on Power Apparatus and Systems*, vol. PAS-104, no. 6, pp. 1519–1525, June 1985.

30 IEEE Power System Relay Committee, "Motor Bus Transfer," *IEEE Transactions on Power Delivery*, vol. 8, no. 4, pp. 1747–1758, October 1993.

Protection of Motor-Operated Valves

31 J. D. Kueck, "An Investigation of Magnesium Rotors in Motor-Operated Valve Actuators," *IEEE Transactions on Energy Conversion*, vol. 3, pp. 40–43, March 1988.

32 A. Richards and C. Formica, "Motor Overload Protection for Motor Actuated Valves," *IEEE Transactions on Power Apparatus and Systems*, vol. PAS-100, pp. 43–50, January/February 1981.

33 IEEE Working Group PES-NPEC-SC4.7 Report, "Design Features and Protection of Valve Actuator Motors," no. 90 WM 094-3 EC, 1990 Winter Meeting.

9

Interface of the Nuclear Plant with the Grid

9.1 Preferred Power Supply Safety Function

The PPS is the preferred power supply for the Nuclear Generating Station Class 1E systems, and the PPS circuits are used during all modes of operation to supply power to the Class 1E and non-Class 1E buses of the plant, but the safety function of the PPS is to provide electric power, as required, for the safe shutdown of the station and for the operation of safety systems. Guidance is provided in IEEE 336 [14], IEEE 338 [15], IEEE 384 [16].

While the PPS is very important to the safety of the plant, it is nevertheless classified as non-Class 1E. Because of its importance to safety, to the extent possible various requirements for redundancy, independence, and separation, which are associated with Class 1E installations still apply. (See NRC Grid Reliability study [6].)

Requirements of the PPS

General

The electrical path of the PPS to the Class 1E buses consists of the following elements:

- the connections from the Class 1E system to the switchyard,
- the switchyard,
- the transmission lines connecting the switchyard to the grid, and
- the transmission grid.

The connections from the Class 1E to the switchyard may consist of overhead, or underground cables, or buses. The connections are classified as non-Class 1E, from the Class 1E circuit breaker terminals. The Class 1E circuit breaker marks the boundary between Class 1E and non-1E equipment. The connections to the switchyard include one or more transformers, which elevate the voltage from the Class 1E bus to the switchyard voltage. These transformers are often of the self-regulating type, incorporating load tap changers, which help

Electrical Systems for Nuclear Power Plants, First Edition. Omar S. Mazzoni.

in providing acceptable voltages to the Class 1E bus. Guidance for transformer protection is provided in [10], [11], [12], and [21].

The switchyard may consist of one or more separate units at different voltage levels, for example, 230 and 500 kV. The switchyard includes transformers that provide for the voltage to be elevated to the grid transmission voltage level. Various schemes are utilized for the switchyard design. A widely utilized scheme is the one called "breaker and a half," which consists of two main busses connected together by a series line up connection of three breakers. The connection of lines and the main generator transformer are made at the points between two breakers. The normal operation of this type of switchyard design is with all breakers continuously closed, which allows for isolation of a bus fault without service interruption. As a result, this scheme design has high reliability and is therefore preferred for the PPS.

The switchyard is generally designed as an outdoor installation and obviously exposed to weather conditions, including tornados, hurricanes, and typhoons. Sulfur hexafluoride (SF6)-insulated stations, also known as gas-insulated substations, could be utilized for the switchyard, which would provide greater protection against weather. SF6 is an inorganic, colorless, odorless, nonflammable gas, which is an excellent electrical insulator. However, SF6-insulated switchyards demand a great financial expense, which could only be justified where environmental conditions are extremely limiting, such as exposure to severe contamination from sand and sea salt. In any case, the protection provided by an SF6 installation would be limited, as the transmission lines (which are part of the PPS) are obviously exposed to the weather. It is important to realize that, despite its exposure to the weather, the typical outdoor switchyard has many accumulated years of reliable service. (See Reliability Council Planning Standard [7].)

The transmission lines connecting to the switchyard are generally in the range of 230–500 kV. Because the transmission lines are exposed to weather conditions, including tornados, hurricanes, and typhoons, they should be designed such that they have as much physical separation as possible. The separation is required to comply with the requirements of 10CFR50 [3], Appendix 17, the PPS should provide at least two separate and independent paths from the grid to the Class 1E buses.

Status indication (open-close) of PPS circuit-interrupting devices need to be provided in the main control room. Undervoltage alarms should be provided in the main control room for the DC systems associated with the PPS circuit-interrupting devices. An alarm to sense voltage degradation should be provided in the main control room. Also, indication of the PPS voltage and frequency should be provided in the main control room.

Availability, Capacity, Capability, and Independence

A minimum of two circuits from the transmission system to the Class 1E power system should be designed to be available during start-up and normal operation to meet accident, postaccident, and safe shutdown requirements in accordance with the following guidelines:

- The two circuits are connected by way of the non-Class 1E distribution system to the redundant Class 1E busses. The non-Class 1E connections should be designed to not compromise the PPS independence requirements.
- At a minimum, one PPS circuit should be immediately and automatically available to provide power to the Class 1E following a design basis accident. It is preferred to have the two PPS circuits automatically and immediately available, but Nuclear Regulatory Commission (NRC) regulations accept having the second PPS circuit available within a time period demonstrated to be adequate by the safety analysis of the station. Some power plant designs utilize the backfeed option through the main step-up transformer as the second feed to the PPS. This option requires disconnecting the generator links and eliminates the ground protection, which was provided by the generator neutral. This option is acceptable if it is provided with a separate ground detection system for the low voltage side of the step-up transformer.

Each circuit of the PPS should be designed to provide sufficient capacity and capability to power equipment that is required to supply the maximum expected coincident Class 1E and non-Class 1E loads.

The PPS circuits to the Class 1E power system should be physically independent, designed, and located to minimize (to the extent practical) the likelihood of simultaneous failure of both PPS circuits under operating, postulated accident, and environmental conditions. For example, it is preferable to design separate transformers for the connection to the switchyard from the redundant Class 1E buses.

Likewise, to reduce the likelihood of simultaneous failure of both PPS circuits, the PPS control circuits should be physically independent from each other and from the PPS power circuits.

If a common takeoff structure is used for both PPS circuits, analysis should be performed to confirm that the reliability of the common takeoff structure is not lower than the overall reliability of the balance of the PPS circuits.

General Design Criterion 2, "Design Bases for Protection Against Natural Phenomena," set forth in Appendix A to 10 CFR Part 50 [3] requires, in part, that structures, systems, and components (SSCs) that are important to safety in nuclear power plants must be designed to withstand natural phenomena. The design bases for these SSCs must reflect (1) appropriate consideration of the most severe of the natural phenomena that have been historically reported for the site and surrounding area, with sufficient margin for limited data, quantity, and period of time in which the historical data have been accumulated; (2) appropriate combinations of the effects of normal and accident conditions with the effects of the natural phenomena; and (3) the importance of the safety functions to be performed. The NRC summarized the risks associated with Grid reliability in [6].

In accordance with the above requirements, the design basis should include consideration of environmental conditions to which the PPS may be subjected at the particular site location, including:

- lightning,
- ice and/or snow,
- temperature variations,
- flood,
- tornados,
- hurricanes, and
- wind.

In addition, the design should include identification of transmission system steady state, and transient conditions to which the PPS may be subjected, including

- switching and lightning surges,
- voltage ranges—maximum and minimum—for heavy and light load conditions,
- voltage transient response,
- frequency variation and transient response,
- stability limits,
- contingency planning for faults,
- coordination of protective relaying,
- grounding.

The design basis should include the capability of periodic testing of protective relays, breakers, transformers, automatic transfer schemes, buses, ground grid, control and instrumentation circuits and devices, DC system, and communication systems. Protection of communication lines is of vital importance, IEEE 487 [18] provides guidance in this area.

Lightning and Surge Protection

Lightning strikes to the transmission lines may originate high voltage surges that may travel into the plant systems connected to the transmission network. Lightning protection guidance is provided in IEEE Std C62.2 [22], IEEE Std C62.41 [23], and IEEE Std C62.45 [24].

Fast transient overvoltages generated by lightning discharges can cause equipment damage, system malfunctions, or power interruptions at nuclear power generating plants if the plants are not adequately protected against such conditions. Power system transients are analyzed in [9]. However, adequate equipment and system design can greatly reduce or alleviate the adverse consequences of abnormal voltage disturbances (see NUREG/CR-6866) [1].

Reg Guide 1.20, "Guidelines for Lightning Protection of Nuclear Power Plants" [2] establishes the requirements for lightning protection. The scope of the guidance includes protection of

- the power plant and relevant ancillary facilities, with the boundary beginning at the service entrance of buildings;
- the plant switchyard;
- the electrical distribution system;
- safety-related instrumentation and control (I&C) systems;
- communications; and
- other important equipment in remote ancillary facilities that could impact safety.

The scope includes signal lines, communication lines, and power lines, as well as testing and maintenance.

Secondary effects of lightning discharges on safety-related I&C systems, such as low-level power surges and electromagnetic and radio-frequency interference (EMI/RFI) are covered in Regulatory Guide 1.180, "Guidelines for Evaluating Electromagnetic and Radio-Frequency Interference in Safety-Related Instrumentation and Control Systems." Regulatory Guide 1.180, [26] which the NRC issued in January 2000 and revised in October 2003, addresses design, installation, and testing practices for dealing with the effects of EMI/RFI and power surges on safety-related I&C systems.

Reliability
The PPS transmission system reliability should be determined using analytical or modeling techniques to ensure that availability requirements are met with future-planned transmission and generation facilities considered.

9.2   Interface between the Nuclear Plant and the Grid

The transmission grid operator and the nuclear plant operator should establish the following criteria for interface (see IEEE 765):

- the number of transmission circuits required to be in service for normal, start-up, shutdown, and accident conditions,
- the MW and megavolt ampere reactive (MVAR) demand of the nuclear power generating station on the PPS for normal, start-up, and accident scenarios,
- the minimum and maximum voltage needed to support the nuclear power generating station during start-up, shutdown, and accident scenarios,
- identification of operational and initiating events that cannot compromise the requirements of the preferred power system,

- switchyard normal voltage schedule,
- minimum and maximum MW and MVAR generation limits,
- grid frequency and stability operating limits,
- nuclear power generating station unique requirements such as independence, environmental factors, and single failure criteria, and
- required actions for prompt restoration of power in the event of loss of all preferred power to the nuclear power generating station.

9.3 Transmission Line and Switchyard Protective Relaying

Protective systems are provided to minimize the probability of loss of the PPS under fault conditions. High voltage (500, 380 kV) transmission lines are generally protected with redundant high speed relay schemes with reclosing and communication equipment to minimize line outages. Two fully redundant protective relaying systems are normally provided: primary and backup relay systems. Each one of these two channels of relaying is connected to one of the two trip coils provided at every breaker. In addition, two DC systems are provided at the switchyard. Each one to the DC systems is dedicated to one of the protective relaying schemes, that is the primary and backup channels.

The switchyard DC system channels are generally segregated from each other, separated by physical distance and often in separate rooms. The control, power, and instrumentation wiring for each channel is separated from the redundant channel by different protecting enclosures and by distance.

A sequence of events recorder (SER) is normally provided for the switchyard systems. The SER will provide valuable information for the study of system faults and other perturbations. It is capable of tracking events and providing event graphing prior, during, and after a system perturbation. The protective relays, DC system equipment, and other control and instrumentation are normally housed in a separate control house located inside the switchyard.

Breakers are provided with a breaker failure scheme that isolates a breaker that fails to trip due to a malfunction.

High voltage (500, 765, 345, 230 kV) transmission lines are generally protected with redundant high-speed relay schemes with reclosing and communication equipment to minimize line outages. High voltage switchyard buses have redundant bus differential protection using separate and independent current and control circuits. Generating unit tie lines and auxiliary transformer underground cable circuits are protected with redundant high-speed relay schemes. Transformers are protected with differential and overcurrent relay schemes. Protection of power transformers should be in accordance with IEEE C37.91 [21].

9.4   Connections of the PPS to the Class 1E Systems

The connection between the preferred power supply and the Class 1E power systems should be made at the input terminals of the Class 1E circuit breaker. (The class 1E circuit breaker is located in safety class structures in accordance with IEEE 308 [13].)

When bus transfer schemes are used, operation of the Class 1E breakers or non-Class 1E breakers to transfer from one PPS source to another may be accomplished by use of the following:

- manual transfer schemes,
- automatic live-bus transfer, and
- automatic dead-bus transfer.

Where automatic bus transfers are used, the adequacy of the transfer schemes needs to be demonstrated in terms of avoiding out of synchronism conditions that may impose undue stresses on the connected Class 1E motors.

The transfer of buses is utilized for plant start-up and shutdown, as well as for maintaining continuity of service.

For plant start-up, the PPS is utilized to provide power to both the Class 1E and non-1E buses. After the main generator is synchronized and loaded to about 30% power, the non-1E buses are manually transferred to the main generator. This manual transfer is performed by closing of the non-1E breaker onto the PPS. Normally, the closing of the non-1E breaker provides for automatic tripping of the 1E source at the non-1E bus, such that the period of paralleling of the PPS with the main generator source is very brief (in the order of a few seconds). The paralleling of the two sources is brief to preclude concerns with the short circuit interrupting and momentary capabilities of breakers involved in the paralleling. An inverse scheme of bus transfer occurs for plant shutdown.

Automatic live transfer is performed under certain conditions to provide continuity of power in case of a bus fault. The automatic live transfer is by either breaker auxiliary contacts or by protective relaying or by both. Additional information on bas transfers is provided in Chapter 8.

9.5   Switchyard Grounding

The switchyard grounding function is to provide a safe path for a ground fault current to ground and to limit the effect of ground potentials gradients to such voltage and current levels that will not endanger the safety of personnel under normal and fault conditions.

Computer programs are available to design and verify the adequacy of ground grids. For a first approximation, to calculate the value of the maximum ground fault current, formula 38 of IEEE 80 [25] can be used.

9.6 Switchyard and Transmission Line Surveillance and Testing

Transmission lines are inspected via an aerial inspection program approximately twice per year. The inspection focuses on such items as right-of-way encroachment, vegetation management, conductor and line hardware condition, and the condition of supporting structures.

Routine switchyard inspection activities include, but are not necessarily limited to, the following:

- daily transformer inspections,
- periodic inspections of circuit breakers and batteries,
- quarterly infrared scans,
- semiannual infrared scans (relay panels),
- semiannual inspection of substation equipment,
- annual infrared scans, and
- annual corona camera scan.

Routine switchyard testing activities include, but are not necessarily limited to, the following:

- biennial circuit breaker profile or timing tests,
- biennial 500 kV relay testing,
- triennial 230 kV relay testing,
- 5-year battery discharge testing,
- 8-year potential transformer testing,
- 8-year ground grid testing,
- 10-year capacitor voltage transformer testing,
- 10-year arrester testing, and
- 10-year wave trap testing.

9.7 Effect of PPS Voltage Degradation on the Class 1E Bus

PPS voltage degradation should be detectable at the Class 1E bus to which the PPS source is connected. Selection of degraded voltage and time delay set points should be in accordance with IEEE Std 741 [19]. A PPS voltage degradation condition should be alarmed in the control room. The affected PPS circuits should be automatically disconnected from the Class 1E buses on sensing PPS degradation to a low-voltage condition below the minimum value that will assure proper operation of all electrical loads required for mitigation of design basis events.

9.8   Multiunit Considerations

Sharing of PPSs

At a multiunit station, the PPS may be shared between two or more units. PPS circuits that are shared among units should be capable of simultaneously supplying all loads required for each design mode of operation of the units.

Status indication of the shared PPS circuit-interrupting devices should be located in the main control room of each unit that is sharing these facilities.

Protective systems of shared PPSs should be provided to minimize the probability of loss of the shared PPS. Protective systems should be designed to minimize the impact of events in one unit on the other unit(s).

The design basis should specifically address the degree of independence of the PPSs of each unit. The design basis should address shared major equipment, the impact of their loss on each unit, interactions among units, and expected operational occurrences and disturbances on each unit.

The capability to monitor the availability of the PPS in the control room of each unit should be provided. The ability to disconnect one unit from its PPS should not affect the PPS availability to the other unit(s). When the PPS is disconnected from the transmission system, indication should be provided at the control board of each unit. If controls are located at multiple points, interlocks should be provided to ensure that the control action will be from only one location at any given time, with appropriate indications as to which location has access to control.

9.9   Considerations for PPS Reliability in a Deregulated Environment

As the electric utility industry moves toward deregulation, the reliability of the preferred power system to a nuclear power generating station must be maintained. The purpose of this section is to identify the design bases and other interface requirements that should be considered and acknowledged between a nuclear power generating station and independent system operators or other organizations not corporately aligned with the utility. The licensing bases for each nuclear power generating station determines the specific functional requirements and limits of the preferred power system with respect to the nuclear facility. This section identifies areas in a suggested format that need to be considered in the preservation of the reliability of the preferred power system in a deregulated environment.

The need for a reliable preferred power system, with the capacity and capability to function in support of a nuclear power generating station, requires the close cooperation and interface of the utility's transmission, system planning, and nuclear organizations. In a regulated environment, this cooperation is facilitated by interface documents, procedures, and a common management structure. In a deregulated environment, organizations that are required to operate and maintain the grid and preferred power system may be independent from the nuclear power generating station and operate on an independent, commercial platform. The interface between the nuclear power generating station and these organizations may become a matter of contract rather than procedure.

Furthermore, in order for a nuclear power generating station to operate, it must meet its technical specifications. This demands that clear and unambiguous requirements exist between the nuclear power generating station and the preferred power provider. (See [7] for interface with the grid.)

IEEE 765 [20] provides a suggested Preferred Power System Interface Agreement with the objective to provide a structure of an agreement document and the suggested contents to ensure that all parties understand their responsibilities in the design, maintenance, and operation of the off-site power system.

Transmission System Studies

Transmission system studies (TSS) should be performed to demonstrate that the PPS is not degraded below a level consistent with the availability goals of the plant as a result of contingencies such as any nonsimultaneous events of the following:

- loss of the local nuclear power generating unit,
- loss of the largest generating unit,
- loss of the largest transmission circuit or intertie, and
- loss of the largest load.

System studies should be reviewed periodically for transmission, generation, and system load changes to ensure that the design basis of the PPS remains valid for all reasonably expected system conditions. Guidance for power system studies is given in [17].

TSSs are typically performed by a transmission planning/operations organization, which is usually separate from the nuclear organization. Good communications across organizations is critical when it comes to defining the plant/grid interfaces. IEEE 765 stresses the need for a strong interface between the transmission system owner/operator and the plant. Important interfaces are input data, modeling methods, design and licensing bases, and interpretation of TSSs. The plant provides information on the worst case station service loading for the different plant operating conditions, including any load transfers or additions that may result following a plant trip. The transmission organization should take that station service loading and consider transmission contingencies under different transmission system loading conditions and configurations. Many times the information supplied by the plant contains contingencies and when coupled with the transmission system contingencies; the results of the studies may show that little or no margin exists. This is primarily due to too many unrelated simultaneous contingencies being evaluated.

TSSs include assumptions concerning contingencies (failures) to be considered in system studies.

Study Methods

The study of PPS adequacy is often performed in two different environments: (a) the traditional transmission planning study validates the overall design of the transmission system for the current year and usually a future year look-ahead and (b) transmission operating studies are performed generally for an upcoming year and are used to identify any transmission operating restrictions and/or compensatory measures for expected system loading scenarios and system configurations. Alternately, real-time operational studies may be performed to ensure that the transmission system is being operated consistent with the philosophy reflected in the transmission planning study.

TSSs performed for nuclear plant operating license compliance are intended to demonstrate that the transmission system, including all connected generation, is designed and constructed with enough capacity and capability to provide the nuclear plant's station service load needed to shut down and mitigate a postulated accident. This objective is accomplished by providing sufficient flexibility in the transmission system such that a transmission element outage does not jeopardize the capability and capacity of the PPS. These TSSs should also consider the worst case bulk system power transfers.

Criteria

TSSs that are input to nuclear plant voltage analyses should use include consideration of specific design basis limits such as allowable low voltage limits and per cent voltage drop at the nuclear power plant should be provided to the PSS transmission operator (TO). Periodically, these design limits should be verified by the nuclear power plant to be correctly implemented by the TO. TSSs may need to be revised when the input data, such as the plant's load requirements or the transmission system configuration changes.

Minimum required posttrip grid voltages should be established such that sufficient voltage would be supplied to the onsite plant distribution system in the event of a postulated design basis event at the plant and operating/shutdown of other nuclear operating units at the site as required by the licensing bases.

Concurrently, TSSs should ensure that the requirements of nuclear plants are met under the range of transmission operating conditions anticipated, assuming that transmission elements may be unavailable, such as during periods of transmission system or neighboring generator maintenance.

The term "availability goals" is meant to reflect the ability of the PPS to meet the station voltage requirements of the connected loads.

The above practices are intended to facilitate the design and operation of the transmission system to minimize the anticipated occurrences of grid conditions resulting in unacceptable offsite power voltages at nuclear plant switchyards.

Maximum acceptable grid voltage should be established based on light in-house plant loading, such as during refueling.

The transmission provider/system operator should have provisions in place to

1. recognize when conditions exist which would result in unacceptable offsite power voltages at nuclear plant switchyards,
2. promptly notify nuclear plant operating personnel of such actual or anticipated conditions, and
3. restore conditions to an acceptable level, if possible.

9.10 Alternate AC Source

To cope with station blackout (SBO) conditions, many US nuclear plants have installed alternate AC (Aac) current power sources that are available and located at, or nearby nuclear power. The AAC is not part of the PPS, it is just a means to provide ac power under SBO conditions. For detail information on SBO, see chapter 10.

The AAC source is designed so that it has a minimum potential for common mode failures with off-site power or on-site emergency ac power sources. It is designed to avoid vulnerability to weather-related events or single failure that could render inoperable, both the AAC power source and the PPS. As an example, connection to a common switchyard for both the AAC and the PPS should be avoided.

The AAC should be available in a time period consistent with the SBO analysis. It should have sufficient capacity and reliability to operate all systems required for both coping with a SBO and for the time needed to bring the plant to, and maintain it in a safe shutdown condition, under nondesign basis accident.

9.11 Study of Recent Events

Actual Case of Transmission Grid Event Affecting a Nuclear Plant

Bulletin No. 12-086 [8] from the **U.S. NUCLEAR REGULATORY COMMIS-SION**

July 27, 2012

NRC ISSUES BULLETIN ON POTENTIAL DESIGN VULNERABILITY IN ELECTRIC POWER SYSTEMS AT NUCLEAR POWER PLANTS

The Nuclear Regulatory Commission has issued a Bulletin to all nuclear power plant licensees requesting information about their electric power system designs and alerting them to a potential design vulnerability that could affect the operation of key safety equipment.

On Jan. 30, Byron Station, Unit 2, shut down automatically due to unbalanced voltage entering the onsite power distribution system from the transmission network. The plant's electric power system's protection scheme was not designed to sense the loss of one of three power phases and automatically trip circuits to isolate the degraded outside power source and switch to emergency backup power. Plant operators diagnosed the problem eight minutes later and manually tripped the necessary circuits.

The degraded offsite power source potentially could have damaged the plant's emergency core cooling system. NRC regulations and plant technical specifications require reliable off-site and onsite power systems with sufficient capacity and capability to operate safety-related systems. Therefore, the NRC staff is seeking information to verify compliance with existing regulations and plant-specific licensing bases. (See IN 95-37 [4] and IN 2000-06 [5].) The staff will use the information it receives to inform the Commission and determine if further regulatory action is needed.

The NRC issued Information Notice 2012-03 on March 1 to inform licensees of recent experience involving loss of one of three phases of the offsite power circuit, including the Byron event. The current Bulletin requests licensees to provide information on their electric system designs within 90 days. It applies to the 104 currently operating commercial power reactors and the four combined licenses for new reactors issued earlier this year. It does not apply to licensees who have permanently ceased operation and certified that the fuel has been removed from the reactor vessel.

Questions and Problems

9.1 *Application of Voltage Regulating Transformer*

Control of power into a distribution network by use of a voltage regulating transformer.

This problem can is to be solved by using the method of a circulating current flowing in the loop formed by parallel branches of a distribution circuit. The voltage drop caused by the circulating current should equal the desired voltage compensation.

Two buses a and b are connected to each other through impedances $X_1 = 0.08$ and $X_2 = 0.12$ PU in parallel. If Va = 1.05(10°) and Vb = 1.0(0°) PU, assume a regulating transformer is inserted in series with X_2 such that no vars flow into bus from the branch whose reactance is X_1. Use the circulating current method and neglect the reactance of the regulating transformer and further assume that P and Q of the load and Vb remain constant.

This problem can be solved by using the method of a circulating current flowing in the loop formed by parallel branches of a distribution circuit. The voltage drop caused by the circulating current should equal the desired voltage compensation.

9.2 *Voltage Surge Analysis*

At a nuclear power generating plant a lightning strike to a transmission line generates a surge that travels down to the line to the plant switchyard and impinges upon the circuit connecting the switchyard and the Class 1E bus. The circuit from the switchyard to the Class 1E bus is made up of distribution line connected to an underground cable.

The surge has a square front of 600 kV. The characteristics of the line and cable are as follows:

Overhead line $Z_o = 280\,\Omega$, $v = 980$ ft/μs

Cable $Z_o = 50\,\Omega$, $v = 400$ ft/μs

The end of the cable is connected to a load with characteristic impedance of 1000 Ω. The cable is 2000 ft long.

Calculate
a) the value of the surge that enters the cable,
b) initial current in load,
c) the reflected surge at the cable/line joint after the first reflection from the load,
d) current flowing in the load 18 μs after the surge reaches the junction of the line/cable (time zero), and
e) voltage at line/cable junction immediately after the first reflected wave has returned.

9.3 *Switchyard Ground Resistance*

Nuclear power generating station switchyard resistance to ground

Utilizing Equation 38 of IEEE 80, calculate the approximate value of the maximum ground current for a switchyard with the following data:
Area: 1300 m^2
Grid resistivity: 2000 Ω-m

9.4 *PPS Available Circuits from the Grid*

A nuclear plant utilizes the backfeed option through the main step-up transformer as the second feed to the PPS. This option requires disconnecting the generator links and eliminates the ground protection which was provided by the generator neutral. This option is acceptable if it is provided with a separate ground detection system for the low voltage side of the step-up transformer.

a) Provide a schematic diagram of an acceptable ground detection system utilizing open delta potential transformers.

b) Calculate the voltage ground detection relay under a phase to ground fault

9.5 *Differential Protection for Transformer for Connection between the PPS and the Class 1E Bus*

A 50-MVA transformer bank, wye-grounded connected to a 230-kV bus, and delta to a 4.16-kV bus, supplies power to the 13.8-kV system. Transformer breakers are available on both sides of the bank with 300:5 (115 kV side) and 2200:5 (13.8 kV side) current transformers. To ground the 13.8-kV system, a 1200-kVA zigzag transformer has been connected between the power transformer and the 13.8-kV bus and within the differential zone. For this arrangement:

Connect three two-restraint type transformer differential relays to protect the 50-MVA bank using the two sets of CTs on the breakers.

Only these are available.

Show all connections in a three-line diagram.

9.6 *Current Grid Events Affecting Nuclear Stations*

Referring to NRC Bulletin No. 12-086 [8], provide a suggested approach to help solving the issue.

References

1 NUREG/CR-6866, "Technical Basis for Regulatory Guidance on Lightning Protection in Nuclear Power Plants," July 2005.
2 Regulatory Guide 1.20, "Guidelines for Lightning Protection of Nuclear Power Plants."
3 General Design Criterion 2, "Design Bases for Protection Against Natural Phenomena," set forth in Appendix A to 10 CFR Part 50.
4 NRC Information Notice IN 95-37, "Inadequate Offsite Power System Voltages during Design Basis Events."
5 NRC Information Notice IN 2000-06, "Offsite Power Voltage Inadequacies."
6 NRC Regulatory Risk Issue Summary 2004-05, "Grid Reliability and the Impact on Plant Risk and the Operability of Offsite Power."
7 North American Electric Reliability Council (NERC) Planning Standard I.A, "Transmission Systems."
8 NRC Bulletin No. 12-086, July 27, 2012, "Potential Design Vulnerability in Electric Power Systems at Nuclear Power Plants."

9 A. Greenwood, *Electrical Transients in Power Systems*, John Wiley & Sons, Inc., New York, 1991.

10 J. L. Blackburn, *Protective Relaying Principles and Applications*, Marcel Decker, Inc., New York, 1987.

11 IEEE 242, "Recommended Practice for Protection and Coordination of Industrial and Commercial Power Systems."

12 IEEE Std 141, "IEEE Recommended Practice for Electric Power Distribution for Industrial Plants (ANSI)."

13 IEEE Std 308, "IEEE Standard Criteria for Class 1E Power Systems for Nuclear Power Generating Stations (ANSI)."

14 IEEE Std 336, "IEEE Standard Installation, Inspection, and Testing Requirements for Power, Instrumentation, and Control Equipment at Nuclear Facilities (ANSI)."

15 IEEE Std 338, "IEEE Standard Criteria for the Periodic Surveillance Testing of Nuclear Power Generating Station Safety Systems (ANSI)."

16 IEEE Std 384, "IEEE Standard Criteria for Independence of Class 1E Equipment and Circuits (ANSI)."

17 IEEE 399, "Recommended Practice for Power System Analysis."

18 IEEE Std 487, "IEEE Recommended Practice for the Protection of Wire-Line Communication Facilities Serving Electric Power Stations (ANSI)."

19 IEEE Std 741, "IEEE Standard Criteria for the Protection of Class 1E Power Systems."

20 IEEE Std 765, "IEEE Standard for Preferred Power Supply (PPS) for Nuclear Power Generating Stations (ANSI)."

21 IEEE Std C37.91, "IEEE Guide for Protective Relay Applications to Power Transformers (ANSI)."

22 IEEE Std C62.2, "IEEE Guide for Application of Gapped Silicon-Carbide Surge Arresters for Alternating Current Systems (ANSI)."

23 IEEE Std C62.41, "IEEE Recommended Practice for Surge Voltages in Low-Voltage AC Power Circuits (ANSI)."

24 IEEE Std C62.45, "IEEE Guide on Surge Testing for Equipment Connected to Low-Voltage AC Power Circuits (ANSI)."

25 IEEE Std 80, "AC Substation Grounding."

26 NRC Regulatory Guide 1.180, "Guidelines for Evaluating Electromagnetic and Radio-Frequency Interference in Safety-Related Instrumentation and Control Systems."

10

Station Blackout: Issues and Regulations

10.1 Introduction

The term "station blackout" (SBO) refers to the complete loss of alternating current (ac) electric power to the essential and nonessential switchgear buses in a nuclear power plant (NPP). An SBO, therefore, involves the loss of the off-site electric power system (the preferred power supply) concurrent with a main generator trip, and unavailability of the emergency alternating current (EAC) power system (typically emergency diesel generators [EDGs]). Because many safety systems necessary for reactor core decay heat removal depend on ac power, an SBO could result in a severe core damage accident.

An SBO is therefore, much more severe than a loss of off-site power (LOOP), which assumes only a loss of off-site power. Under a LOOP, nevertheless, it is also assumed that the most sever DBA (design basis accident) occurs concurrently.

Under an SBO, it is assumed that the on-site safety-related DC systems are operable, as well as the ac power produced by inverters fed by the DC batteries. Also assumed under an SBO is that the plant does not undergo a DBA. An earthquake (as well as similar strong environmental perturbations), a simultaneous DSA, and a SBO are considered to be mutually exclusive events by the nuclear generating industry. The rational for this assumption lies in the very low probability for the simultaneous occurrence of an SBO event and either a DBA or severe environmental conditions. This assumes that the design of the plant is up to all applicable standards and that the operation of the plant includes all required testing and surveillance.

The risk evaluation of an SBO involves the estimation of the likelihood and duration of the loss of all ac power and the potential for severe core damage after a loss of all ac power.

SBO was *not* considered in the early design of plants, and due to several incidents of SBOs at operating plants, in 1980 the Nuclear Regulatory Commission (NRC)-designated SBO as an unresolved safety issue (USI A-44). The agency documented the findings of the technical studies completed for USI A-44 in

Electrical Systems for Nuclear Power Plants, First Edition. Omar S. Mazzoni.
© 2019 by The Institute of Electrical and Electronic Engineers, Inc. Published 2019 by John Wiley & Sons, Inc.

NUREG-1032 [20]. In June 1988, the NRC resolved USI A-44 with the publication of a new rule under 10 CFR 50.63 (53 FR 23203) and an accompanying regulatory guide (RG) (RG 1.155).

Concurrently, the Nuclear Management and Resources Council (NUMARC) (now the Nuclear Energy Institute (NEI)) developed NUMARC-8700, Revision 0 (Ref. 24), which RG 1.155 [5] endorses with certain exceptions. Table 1 of RG 1.155 provides a cross-reference to NUMARC-8700, [24] Revision 0, and it notes when the RG takes precedence.

The NRC determined that information presented in the safety analysis report (SAR) needs to be sufficient to support the conclusion that the plant is capable of withstanding, and recovering from, a complete loss of ac electric power to the essential and nonessential switchgear buses for a specified period of time.

10.2 Regulations Relating to SBO Requirements

There are several regulations that provide a technical rationale for application SBO requirements. A plant needs to ascertain whether or not it complies with the following regulations. Adequacy of the design allows for compliance with the regulations, guidelines for design adequacy are provided in NRC and industry standards, such as [8] and [9]. Applicable regulatory guides are [4], [6], and [7].

Compliance with GDC 17

GDC 17 requires that on-site and off-site electrical power be provided to facilitate the functioning of structures, systems, and components important to safety [2]. Each electric power system, assuming the other system is not functioning, must provide sufficient capacity and capability to ensure that specified acceptable fuel design limits and the design conditions of the reactor coolant pressure boundary are not exceeded as a result of anticipated operational occurrences and that the core is cooled and containment integrity and other vital functions are maintained in the event of postulated accidents. GDC 17 also requires the inclusion of provisions to minimize the probability of losing electric power from any of the remaining supplies as a result of, or coincident with, the loss of power generated by the nuclear power unit, the loss of power from the transmission network, or the loss of power from the on-site electric power supplies. Meeting the requirements of GDC 17 provides assurance that a reliable electric power supply will be provided for all facility operating modes, including anticipated operational occurrences and DBAs, to permit the performance of safety functions and other vital functions, even in the event of a single failure. The NRC has issued numerous documents including Information Notices and Generic Letters that addressed the reliability of the power grid, among which are [10], [11], [13], [14], [15], [16], [21], and [22].

Compliance with GDC 18

GDC 18 [3] requires that electric power systems important to safety be designed to permit appropriate periodic inspection and testing of key areas and features to assess their continuity and the condition of their components. These systems should be designed to test periodically (a) the operability and functional performance of the components of the systems, such as on-site power sources, relays, switches, and buses, and (b) the operability of the systems as a whole and, under conditions as close to design as practical, the full operation sequence that brings the systems into operation, including operation of applicable portions of the protection system, and the transfer of power among the nuclear power unit, the off-site power system, and the on-site power system. Consequently, the ac power system must provide the capability to perform integral testing on a periodic basis. Meeting the requirements of GDC 18 provides assurance that, when necessary, off-site power systems can be appropriately and unobtrusively accessed for required periodic inspection and testing, enabling verification of important system parameters, performance characteristics, and features and detection of degradation and/or impending failure under controlled conditions.

Compliance with 10 CFR 50.63

10 CFR 50.63 requires that each light water cooled NPP be able to withstand and recover from an SBO (as defined in 10 CFR 50.2). Meeting the requirements of 10 CFR 50.63 provides assurance that the NPP will be able to withstand (cope with) and recover from an SBO and will ensure that core cooling and appropriate containment integrity are maintained.

10.3 Specific SBO Requirements

There are two specific areas of SBO requirements: (1) SBO coping duration and (2) SBO coping capability.

SBO Coping Duration

The SBO coping duration is defined as the time from the onset of an SBO to the time when either off-site (preferred) or on-site (EDGs) ac power is restored to at least one of the safe shutdown buses. The SBO rule requires each plant to specify an SBO coping duration that is justified by an analysis of site- and plant-specific factors that contribute to the likelihood and duration of an SBO.

Because passive plants will not have EAC power sources, applicants for such plants need not evaluate SBO coping duration as long as they are able to demonstrate that the design selected is capable of performing safety-related functions for 72 h. The 72-h approach is consistent with the duration approved by the NRC staff for the AP 1000 design. The review should determine that the selected minimum coping duration conforms to the guidance provided in Section C.3.1 of RG 1.155.

SBO Coping Capability

The review should determine that the capability to achieve and maintain safe shutdown (non-DBA) during an SBO conforms to the guidance provided in Section C.3.2 of RG 1.155 and that appropriate procedures and training have been developed to implement this capability.

10.4 Alternate Alternating Current Power Sources

One acceptable means of complying with the requirements in 10 CFR 50.63 [1] involves the provision of an alternate alternating current (Aac) source of sufficient capacity, capability, and reliability for operation of all systems necessary for coping with SBO and for the time necessary to bring the plant to, and maintain it in, safe shutdown condition (non-DBA) that will be available on a sufficiently timely basis.

For new ALWR (advanced light water reactor) plants, the NRC has established a policy that such plants should have an Aac power source of diverse design and capable of powering at least one complete set of normal shutdown loads. In SECY-94-084 [17] and SECY-95-132 [18], [19] the NRC modified these criteria for ALWRs that use passive safety systems. Specifically, an Aac power source is not necessary for passive plant designs (such as the AP1000) that (a) do not need ac power to perform safety-related functions for 72 h following the onset of an SBO and (b) meet the guidelines in Section C.IV.10 of RG 1.206 regarding regulatory treatment of nonsafety systems.

EDGs in excess of minimum redundancy criteria for NPP on-site power systems, nearby or on-site gas turbine generators, portable or other available compatible diesel generators, or hydrogenerators may serve as Aac power sources. The design should meet the recommendations in Sections C.3.2.5, C.3.3, and C.3.5 and Appendices A and B to RG 1.155. It is acceptable for Aac power sources to be normally used for other purposes, and they do not need to be solely dedicated to use as an Aac power source. However, the requisite procedures and interface agreements need to be in place such that the Aac power source is available in an SBO event.

The Aac power source should be available in a timely manner after the onset of SBO and have provisions to be manually connected to one or all of the redundant safety buses as necessary to power all equipment necessary to achieve and maintain safe shutdown (non-DBA). The time necessary for making this equipment available should not exceed 1 h and should be demonstrated by test. If tests can show the Aac power source to be available in less than 10 min, no coping analysis is needed. Otherwise, a coping analysis should be performed for the duration from the onset of the SBO until the Aac power source or sources are started and lined up to operate all equipment necessary to achieve and maintain safe shutdown. The phrase "available within 10 min of the onset of SBO" means that circuit breakers necessary to bring power to safe shutdown buses can be actuated in the control room within that period of time.

To ensure that the requirements of 10 CFR 50.63 regarding the Aac source are satisfied, the NRC evaluates an applicant's submittal on SBO regarding the following issues:

In accordance with Section C.3.3.5 of RG 1.155, the Aac power source should be capable of supplying power, as necessary, to all loads that are required for safe shutdown (non-DBA) in the event of an SBO at any nuclear unit where the Aac is applied. The Aac power source should have sufficient capacity to operate the systems necessary for coping with an SBO for the time necessary to bring and maintain the plant in a safe shutdown condition. The plant systems, functions, and features discussed in Sections C.2 and C.3.3.1 to C.3.3.4 of RG 1.155 should be appropriately addressed as safe shutdown non-DBA loads (including loads associated with any alternative or added capacity battery charging, water, or air sources to handle SBO). For new applications, the Aac source should be of diverse design (with respect to on-site sources); have adequate capacity, independence, and reliability; and have capability for powering at least one complete set of normal safe shutdown loads. At sites where units share on-site emergency sources, the Aac power sources should have the capacity and capability to ensure that all units can be brought to and maintained in a safe shutdown (non-DBA) condition.

Independence of the Aac power source should be such that the Aac source (or sources) will not adversely affect the preferred power system or its specified functions and will not adversely affect the on-site power system or its specified safety functions. Independence should be provided as follows:

With respect to independence between the Aac power source used for SBO and the preferred and on-site power systems, electrical ties between these systems and the physical arrangement of the interface equipment should minimize the potential for the loss of any system (i.e., preferred, on-site, or Aac) preventing access to any other system and the potential for such a loss to cause further failures in other systems.

An acceptable design should not have the Aac power source normally directly connected to the preferred power system or to the blacked-out unit's on-site

EAC power system. No single point of vulnerability should exist whereby a single active failure or weather-related event could simultaneously fail the Aac and preferred power sources or the Aac and on-site sources. The power sources should have minimum potential for common failure modes.

The Aac components should be physically separated and electrically isolated from safety-related components or equipment, as specified in the separation and isolation criteria applicable to the unit's licensing basis and the guidelines in Appendix A to RG 1.155. Based upon compliance with all relevant independence criteria and guidelines, it should be demonstrated that provisions for the Aac source will not, at any time, adversely affect the functioning of off-site and/or Class 1E on-site power systems. Also, failure of the Aac power components should not adversely affect the Class 1E ac power systems.

Careful examination should be made of the physical arrangement of circuits and incoming source breakers (to the affected Class 1E bus or buses), separation and isolation provisions (control and main power), permissive and interlock schemes proposed for source breakers, source initiation/transfer logic, Class 1E load shedding and sequencing schemes that could affect Aac source ability to power safe shutdown loads, source lockout schemes, and bus lockout schemes.

The Aac power source(s) is not automatically connected to the 1E buses for SBO, but it is manually connected to one or all of the redundant safety buses as necessary. However, the design should inhibit the possibility of reducing required independence between the redundant safety buses.

The control room should monitor the performance of the Aac power source. As a minimum, monitoring should include the voltage, current, frequency, and circuit breaker position.

The Aac source components should be enclosed within structures that conform to the Uniform Building Code. Electrical cables connecting the Aac power source to the shutdown buses are protected against the events that affect the preferred ac power system. Buried cables or other appropriate methods can be used to accomplish this.

Nonsafety-related Aac power source(s) and associated dedicated DC system(s) should meet the quality assurance (QA) guidance in Section 3.5, Appendix A, and Appendix B to RG 1.155.

The Aac power system should be equipped with a dedicated DC power system that is electrically independent from the blacked-out unit's preferred and Class 1E power systems and is of sufficient capability and capacity for operation of DC loads associated with the Aac source for the maximum necessary duration of Aac source operation.

The Aac power system should be equipped with a starting system (and motive energy source for starting) that is independent from the blacked-out unit's preferred and Class 1E ac power systems.

The Aac power system should be provided with a fuel supply that is separate from the fuel supply for the on-site EAC power system. A separate day tank, supplied from a common storage tank, is acceptable if the fuel is sampled and analyzed using methods consistent with applicable standards before its transfer to the day tank.

In case that the Aac power source and an emergency on-site ac power source are identical, procedures should be provided to ensure that active failures of each identical power source will be evaluated for common cause applicability and that corrective action has been taken to reduce subsequent failures.

The Aac power system should be capable of operating during and after an SBO without any support system receiving power from the preferred power supply or the blacked-out unit's EAC power sources. The capability of the Aac to start on demand depends on the availability of the necessary support systems to fulfill their required function. These support systems may need varying combinations of DC or ac power for varying periods to maintain operational readiness. Information Notice (IN) 97-21 [12] discusses two examples of a failure of the Aac to start on demand because of an extended loss of auxiliary electrical power sources.

The portions of the Aac power system subjected to maintenance activities should be tested before returning the Aac power system to service.

Plant-specific technical guidelines and emergency operating procedures should be implemented that identify those actions necessary for placing the Aac power source in service.

The Aac power system should be inspected, maintained, and tested periodically to demonstrate operability and reliability. The reliability of the Aac power system should meet or exceed 95% as determined in accordance with NSAC-108 [23] or equivalent methodology.

Where EDGs are identified as Aac power sources, they should meet the following criteria:

- At single unit sites, any EAC power source(s) in excess of the number required to meet the minimum redundancy criteria (i.e., single failure) for safe shutdown may be assumed to be available. These EAC power sources may be designated as Aac power sources, provided the guidelines identified in Section C.3.3.5 of RG 1.155 are met.
- A single unit that needs one EAC source to place the plant in safe shutdown needs one redundant EAC power source. For SBO, both the EAC power source and the redundant EAC power source are unavailable. An EDG may be designated as an Aac power source only if that EDG is neither the necessary EAC power source nor the redundant EAC power source. Therefore, a single-unit requiring one EAC source for safe shutdown should have at least three EDGs, with one EDG that may be designated as the Aac power source meeting RG 1.155 guidance for Aac power sources.

- At multiunit sites, where the combination of EAC sources exceeds the minimum redundancy criteria (on a per-nuclear-unit basis) for normal safe shutdown of all units, the excess EAC power sources may be used as Aac power sources, provided they meet the Aac power source guidance in Section C.3.3.5 of RG 1.155. If no EAC power source in excess of the minimum redundancy criteria remains, the occurrence of SBO must be assumed for all of the units.
- When an SBO occurs at one unit of a multiunit site, the EAC power source(s) and the redundant EAC power source(s) are unavailable. An SBO on one unit does not assume a concurrent single failure; however, the remaining unit(s) should still meet the normal operating single failure criteria. Therefore, an EDG could be designated as an Aac only if (1) the EDG is neither the necessary EAC power source nor the redundant EAC power source for the unit experiencing the SBO and (2) the EDG is not necessary as an emergency or redundant EAC power source for the remaining units. Where an EDG is used as an Aac, it is desirable that the EDG be connectable to all buses essential for normal safe shutdown. Review of the applicant's station on-site ac power system should determine whether such a capability exists.
- Multiunit sites may not use EDGs with 1-out-of-2 (shared) and 2-out-of-3 (shared) ac power configurations as Aac power sources.
- For EDGs used as an Aac source, the engine support systems should conform with the relevant criteria used to evaluate them under SRP (Standard Review Plan, by NRC) Sections 9.5.4 through 9.5.8.

10.5 Procedures and Training

Procedures and training should include all operator actions necessary to do the following:

A. Cope with the occurrence of an SBO for the specified coping duration during all modes of plant operation and include actions necessary to place Aac power sources in service (if used) and maintain acceptable environmental conditions for equipment necessary to mitigate the event. Procedures developed to cope with an SBO should be integrated with the plant-specific technical guidelines and emergency operating procedures developed using the emergency operating procedure upgrade program established in response to Supplement 1 of NUREG-0737. The task analysis portion of the emergency operating procedure upgrade program should include an analysis of instrumentation adequacy during an SBO.

B. Restore standby (Class 1E) power sources when the EAC power system is unavailable.

C. Restore off-site power sources and use of nearby power sources (which may include nearby or on-site gas turbine generators, portable generators, hydrogenerators, and black start fossil power plants) in the event of a LOOP. As a minimum, the reviewer should consider the following potential causes for a LOOP:

1. grid undervoltage and collapse.

2. weather-induced power loss.

3. preferred power distribution system faults that could result in the loss of normal power to essential switchgear buses. This includes such failures as distribution system hardware, switching and maintenance errors, and lightning-induced surges.

D. Actions necessary to restore normal long-term core cooling/decay heat removal once ac power is restored.

E. The procedure should specify actions necessary to assure that shutdown equipment (including support systems) necessary in an SBO can operate without ac power.

F. The procedure should recognize the importance of decay heat removal systems (auxiliary feedwater, high-pressure coolant injection, high-pressure core spray, reactor core isolation cooling) during the early stages of the event and direct operators to invest appropriate attention to ensuring their continued reliable operation throughout the event.

G. Plant-operating procedures should identify the sources of potential inventory loss and specify actions to prevent or limit significant loss.

H. Plant-operating procedures should ensure the prompt establishment of a flowpath for makeup flow from the condensate storage tank (CST) to the steam generator/nuclear boiler and identify backup sources to the CST in order of intended use. In addition, plant-operating procedures should specify clear criteria for transferring to the next preferred source of water.

I. The procedure should identify individual loads that need to be stripped from the plant DC buses (both Class 1E and non-Class 1E) to conserve DC power.

J. Plant-operating procedures should specify actions to permit appropriate containment isolation and safe shutdown valve operations while ac power is unavailable.

K. Plant-operating procedures should identify the portable lighting necessary for ingress and egress to plant areas containing shutdown or Aac equipment requiring manual operation.

L. Plant-operating procedures should consider the effects of ac power loss on area access, as well as the need to gain entry to other locked areas where remote equipment operation is necessary.

M. Plant-operating procedures should consider the effects of a loss of ac power on communications capabilities, including the potential for a loss of communications with off-site agencies.

N. Plant-operating procedures should consider the loss of heat-tracing effects for equipment necessary to cope with an SBO.

O. To provide assurance that the NPP operator will be kept aware of changes in the plant switchyard and off-site power grid, plant or site procedures should establish appropriate communication protocols between the NPP and its transmission system operator. With regard to SBO, these protocols should aid the operator in determining the following:

   i) The performance of grid-risk-sensitive maintenance activities (such as surveillances, postmaintenance testing, and preventive and corrective maintenance) that could increase the likelihood of an SBO or impact the plant's ability to cope with an SBO, such as out-of-service risk-significant equipment (e.g., an EDG, a battery, a steam-driven pump, an Aac power source)

   ii) The availability of local power sources and transmission paths that could be made available to resupply the plant following a LOOP event. Procedures and training should conform to the guidance in Sections C.1.3, C.2, and C.3.4 of RG 1.155. Procedures and training should address all operator actions necessary to (a) restore EAC power when the EAC power system is unavailable, (b) cope with Aac or battery power on the occurrence of an SBO for the specified coping duration during all modes of plant operation, (c) restore off-site power and use of nearby power sources (which may include such items as nearby or on-site gas turbine generators, portable generators, or hydrogenerators) in the event of a loss of LOOP, and (d) restore normal long-term core cooling/decay heat removal once power is restored.

The communication agreements and protocols between the plant and its transmission system operator should provide assurance that the NPP operator will be kept aware of (a) changes in the plant switchyard and off-site power grid and (b) local power sources and transmission paths that could be made available to resupply the plant following a LOOP.

10.6   QA and Specifications for Nonsafety-Related Equipment

QA activities and specifications for nonsafety-related equipment used to meet the requirements of 10 CFR 50.63 should conform to the recommendations in Section C.3.5 and Appendix A to RG 1.155. The systems and equipment used to meet the requirements of 10 CFR 50.63 should conform to the system and station equipment specification recommendations of Appendix A to RG 1.155. Additionally, the nonsafety equipment installed to meet the SBO rule should not degrade the existing safety-related systems. This is accomplished by ensuring that the nonsafety equipment is as independent as practicable from existing safety-related systems.

10.7 Monitoring of the Grid Condition

Pursuant to 10 CFR 50.65, the NPP operator should know the grid's condition before taking a risk-significant piece of equipment out of service and should monitor the grid for as long as the equipment remains out of service. This provides assurance that grid reliability evaluations are performed before undertaking grid-risk-sensitive maintenance activities (such as surveillances, postmaintenance testing, and preventive and corrective maintenance [6], [7]) under existing or imminent degraded grid reliability conditions that could increase the likelihood of an SBO or impact the plant's ability to cope with an SBO, such as out-of-service risk-significant equipment (e.g., an EDG, a battery, a steam-driven pump, an Aac power source).

Questions and Problems

10.1 *Loss of all Class 1E dc power:*
  a) Considering a *hypothetical case* of a loss of all Class 1E dc power determine how it may occur,
  b) from (a) above, discuss the *likelihood* of a Loss of all Class 1E dc power,
  c) examine the *consequences of the loss of all Class 1E dc power* in terms of the continuing operation of the Class 1E ac equipment, analyzing main systems included in the consequences, and
  d) determine the consequences on an SBO if all Class 1E dc power were to be lost.

10.2 *Loss of PPS and loss of EDGs (SBO)*
  A station has the following SBO coping conditions:
  SBO coping time of 4 h.
  SBO coping capability: Class 1E dc batteries.
  Assume the following:
  • A hurricane hit the plant switchyard and completely destroyed the outdoor installation of the PPS,
  • the two redundant EDGs failed to start which resulted in an SBO event,
  • after 4.5 h, the PPS was restored to service

Determine whether the plant was able to successfully survive the SBO event. Explain your answer.

10.3 Provide the approach a plant needs to take to minimize failure of the PPS.

10.4 Provide the approach a plant needs to take to minimize failure of the EDGs.

10.5 Provide the approach a plant needs to take to minimize failure of the DC system.

10.6 Consider the Fukushima accident in terms of the compliance with SBO requirements.

In the case of Fukushima, the upset environmental conditions (tsunami) disabled the EDGs, the PPS, and the Class 1E dc power system. With the disabling of the DC power system, this created an augmented SBO.

Explain what requirements are necessary to be followed in the plant design to prevent a disaster like Fukushima from developing and from ineffective mitigating actions.

10.7 The addition of the Aac source is generally desirable to aid in coping with an SBO. Nevertheless, the need to connect it to the Class 1E system creates concerns with the possibility of degrading of the Class 1E system.

Describe these concerns.

10.8 Consider a plant site which has an acceptable Aac for mitigating an SBO, and assume that there was a very strong tornado and the EDGs failed to start (SBO condition). Describe the potential for the Aac to be available to mitigate the SBO.

References

1 10 CFR 50.63, "Loss of All Alternating Current Power."
2 10 CFR Part 50, Appendix A, GDC 17, "Electric Power Systems."
3 10 CFR Part 50, Appendix A, GDC 18, "Inspection and Testing of Electric Power Systems."
4 Regulatory Guide 1.9, "Selection, Design, Qualification, and Testing of Emergency Diesel Generator Units Used as Class 1E Onsite Electric Power Systems at Nuclear Power Plants."
5 Regulatory Guide 1.155, "Station Blackout."

6 Regulatory Guide 1.160, "Monitoring the Effectiveness of Maintenance at Nuclear Power Plants."

7 Regulatory Guide 1.182, "Assessing and Managing Risk Before Maintenance Activities at Nuclear Power Plants."

8 IEEE Standard 308, "IEEE Standard Criteria for Class 1E Power Systems for Nuclear Power Generating Stations."

9 IEEE Standard 765, "IEEE Standard for Preferred Power Supply (PPS) for Nuclear Power Generating Stations."

10 Generic Letter 2006-02, "Grid Reliability and the Impact on Plant Risk and the Operability of Offsite Power," February 1, 2006.

11 Information Notice 97-05, "Offsite Notification Capabilities," February 27, 1997.

12 Information Notice 97-21, "Availability of Alternate AC Power Source Designed for Station Blackout Event," April 18, 1997.

13 Information Notice 98-07, "Offsite Power Reliability Challenges from Industry Deregulation," February 27, 1998.

14 Information Notice 2000-06, "Offsite Power Voltage Inadequacies," March 27, 2000.

15 Regulatory Issue Summary 2000-24, "Concerns About Offsite Power Voltage Inadequacies and Grid Reliability Challenges Due to Industry Deregulation," December 21, 2000.

16 Regulatory Issue Summary 2004-05, "Grid Reliability and the Impact on Plant Risk and the Operability of Offsite Power," April 15, 2004.

17 SECY-94-084, "Policy and Technical Issues Associated with the Regulatory Treatment of Non-Safety Systems in Passive Plant Designs," March 28, 1994. Approved in the staff requirements memorandum dated June 30, 1994.

18 SECY-95-132, "Policy and Technical Issues Associated with the Regulatory Treatment of Non-Safety Systems (RTNSS) in Passive Plant Designs." Approved in the staff requirements memorandum dated June 28, 1995.

19 NRC Memorandum from D. Crutchfield to File, Subject: Consolidation of SECY-94-084 and SECY-95-132, July 24, 1995. SECY-94-084 was approved in the staff requirements memorandum dated June 30, 1994. SECY-95-132 was approved in the staff requirements memorandum dated June 28, 1995.

20 NUREG-1032, "Evaluation of Station Blackout Accidents at Nuclear Power Plants," June 1998.

21 NUREG-1784, "Operating Experience Assessment—Effects of Grid Events on Nuclear Power Plant Performance," December 2003.

22 NUREG/CR-6890, "Reevaluation of Station Blackout Risk at Nuclear Power Plants - Analysis of Loss of Offsite Power Events: 1986-2004," December 2005.

23 NSAC-108, "The Reliability of Emergency Diesel at US Nuclear Power Plants," September 1986.

24 NUMARC-8700, "Guidelines and Technical Bases for NUMARC Initiatives Addressing Station Blackout in Light Water Reactors," Revision 0, November 1997.

11

Review of Electric Power Calculations

11.1 Introduction

This chapter reviews some of the more important studies and calculations related to nuclear plant electrical systems.

Other important studies are treated in the references, as follows:

- power systems analysis and design: IEEE 141 [2], IEEE 142 [3], IEEE 242 [4], IEEE 308 [5], IEEE 399 [6], IEEE C37.91 [8], and IEEE 765 [7]
- power system protection: Blackburn [1], IEEE 765 [7], IEEE C37.91 [8], IEEE C37.96 [9], IEEE Std C62.92.3 [11], and IEEE C37.102 [10]

11.2 Load and Voltage Calculations

Owing to the vector relationships between voltage, current, resistance, and reactance, voltage drop calculations require a working knowledge of phasors, especially for making exact computations. In general, the voltage drop calculations are performed with sophisticated computer programs, which take into account the plant loading conditions, the nature of the loads, and produce an output that includes the maximum and minimum loading, and the corresponding voltage drop.

Alternatively, for simple cases, most voltage-drop calculations are based on assumed limiting conditions, and approximate formulas are adequate. The vector relationships between the voltage at the beginning of a circuit, the voltage drop in the circuit, and the voltage at the end of the circuit are shown in Figure 11.1, where

Resistance voltage drop $= I\,R \cos \Phi$
Reactance voltage drop $= I\,X \sin \Phi$

From Figure 11.2, the approximate formula for the voltage drop is

$$V_D \text{ (approximate)} = IR\, \cos \Phi + IX \sin \Phi \tag{11.1}$$

Electrical Systems for Nuclear Power Plants, First Edition. Omar S. Mazzoni.

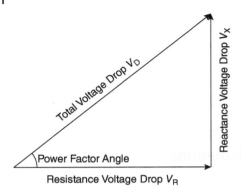

Figure 11.1 Voltage drop.

where V_D is voltage drop in circuit, line-to-neutral, I is current flowing in a conductor, R is line resistance for one conductor, X is line reactance for one conductor, Φ is an angle whose cosine is the load power factor, cos Φ is the load power factor, in decimals, and sin Φ is the Load reactive power factor, in decimals.

The voltage drop V_D obtained from formula (11.1) is the voltage drop in one conductor, one way, commonly called the "line-to-neutral voltage drop." For balanced three-phase systems, the line-to-line voltage drop is computed by multiplying the line-to-neutral voltage drop by the following constants:

Single phase: 2
Three phase: 1.732

In using the voltage drop formula, the line current I is generally the maximum or assumed load current, or the current-carrying capacity of the conductor when limiting conditions need to be evaluated.

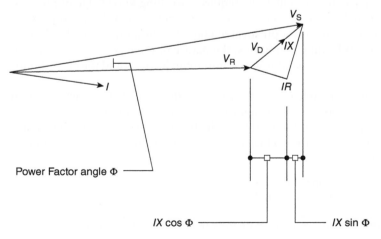

Figure 11.2 Sending end V_S and receiving end V_R.

The resistance R is the ac resistance of the particular conductor used, considering the particular type of raceway in which it is installed. It depends on the size of the conductor (measured in US wire gauge – AWG) for smaller conductors and in thousands of circular mils for larger conductors), the material of conductor (copper or aluminum), the temperature of the conductor generally 90°C (194°F), depending on the conductor rating, for maximum loading, and whether the conductor is installed in a magnetic (steel) or nonmagnetic (aluminum or nonmetallic) raceway.

The reactance X also depends on the size and material of the conductor, whether the raceway is magnetic or nonmagnetic, and on the spacing between the phase conductors of the circuit. The spacing is fixed for a multiconductor cable, but may vary with single conductor cables so that an average value must be used. Reactance occurs because the ac flowing in the conductor causes a magnetic field to buildup and collapse around each conductor in synchronism with the ac. This magnetic field cuts across the conductor itself and the other conductors of the circuit, causing a voltage to be induced into each in the same way that current flowing in the primary of a transformer induces a voltage in the secondary winding. Since the induced voltage is proportional to the rate of change of the magnetic field, which is maximum when the current is passing through zero, the induced voltage will be at maximum when the current is passing through zero or, in vector terminology, the voltage wave is 90° out of phase with the current wave.

Φ is the angle between the load voltage and the load current. cos Φ is the power factor of the load expressed as a decimal and may be used directly in the computation of $IR \cos \Phi$.

$IR \cos \Phi$ is the resistive component of the voltage drop, and it is in phase with the current. $I X \sin \Phi$ is the reactive component of the voltage drop, and it is 90° out of phase with the current. sin Φ is positive when the current lags the voltage (lagging power factor) and negative when the current leads the voltage (leading power factor).

Referring to Figure 11.2, for the receiving end, the voltage expression becomes

$$V_R = [V_S^2 - (IX \sin \Phi - IR \cos \Phi)^2]^{1/2} - (IR \cos \Phi + IX \sin \Phi) \tag{11.2}$$

and for approximate calculations:

$$V_R = V_S - V_D = V_S - (IR \cos \Phi + IX \sin \Phi) \tag{11.3}$$

11.3 Motor Starting Calculations

The voltage criteria, as established for the transmission power flow studies, should be used in the motor starting calculations to determine whether sufficient voltage is available during motor acceleration of the Class 1E loads, without challenging the undervoltage protection and disconnection of the preferred power supply.

Motor starting calculations can be performed using dynamic simulation or static "snap shot" load flows.

Dynamic motor starting analysis will provide more accurate results, but requires detailed information to develop dynamic models for various loads (motors, pumps, etc.). Owing to the complexity of this type of analysis, the use of a computer program capable of dynamic motor simulation is typically required.

When load modeling data are unavailable or when dynamic motor starting simulation is not practical, a static motor starting analysis may be used.

While both types of analysis are acceptable, dynamic motor starting analysis is recommended since it provides more accurate and realistic results. When modeling the ac electrical distribution system, detailed models for larger (medium voltage) motors should be developed. It is acceptable to use bounding, less detailed models for smaller (low voltage) motors, since motor model data are usually not available.

The dynamic motor starting calculation results are evaluated to ensure the motors are properly accelerated, and the undervoltage protection has not been challenged.

In one case, the analysis includes the starting of motors and load sequencing operations are performed, or other critical load changes occur. While this analysis can be a more bounding approach, it can also be overly conservative. This is because the assumptions used for motor acceleration and start times cannot take credit for the inherent interactions found in actual system performance. The static motor starting calculation results are evaluated to ensure that the motors have adequate starting voltage, and the voltage limits of the undervoltage protection have not been challenged.

If any Class 1E motor is found to not accelerate or results indicate unacceptable voltage values to Class 1E loads, design modifications should be made. Examples of design modifications may include one or more of the following:

- strengthen the voltage performance of the distribution system,
- change the undervoltage set points (loss of voltage /loss of voltage),
- change load sequencing times or loading in the EDGS, and
- improve the system impedance by replacing cables and/or transformers.

Calculation of Motor Starting Time

After voltage application to a motor, it begins to accelerate from rest. The motor acceleration will depend on the net accelerating torque. The accelerating torque is obtained by subtracting the motor torque from the load torque at each period of time considered.

Table 11.1 (Extracted from Motor-Load Accelerating Curve, for a Typical Load of Centrifugal Pump)

Period	Speed (%)	T_{motor} (%)	T_{load} (%)	T_{net} (%)	T_{net} (lb-ft^2)
1	0	100	30		
				77.5	2260.4
2	25	120	35		
				100	2916.7
3	50	160	45		
				120	2500.0
4	75	190	65		
				92.5	1822.9
5	95	80	80		

Calculation of the net accelerating torque for a motor, based on values taken from characteristic curves for the load and the motor torques (curves not shown).

$$T_{net} = [T_{motor}(n+1) - T_{load}(n+1)]/2 + (T_{motor}(n) - T_{load}(n)]/2 \tag{11.4}$$

The values of motor torque are displayed in Table 11.1.

IEEE 399 establishes that the motor starting time, as a function of the net accelerating torque is

$$t(s) = [Wk^2 \times (\text{rpm}_2 - \text{rpm}_1) 2\pi]/60g T_n \tag{11.5}$$

where Wk^2 is inertia in lb-ft^2, T_n is net average accelerating torque between rpm$_1$ and rpm$_2$ (lb ft), rpm$_1$, rpm$_2$ are speed at time 1 and time 2, respectively, and $g = 32.2$ ft/s^2.

Short Circuit Calculations

The total rms asymmetrical short circuit current is represented by the following equation:

$$I_{Total} = \text{SQRT} \left[(I_{ac})^2 + (I_{dc})^2 \right] \tag{11.6}$$

where I_{ac} is the ac component of the short-circuit fault, it includes the contributions from motors, generators, and utilities, and is at a maximum during the first half cycle when induction motors are supplying locked rotor current to the fault. I_{dc} is the dc component of the ac fault current and is at a maximum the instant the short circuit occurs.

To cover the worst possible case, the ac fault is assumed to be the maximum possible at the instant the fault occurs. The dc component of the fault current will decay exponentially in accordance with the following equation:

$$(I_{dc}) = \text{SQRT}\,(2) * I_{ac} * e^{(-t)(2\pi f)/(X/R)} \tag{11.7}$$

where

I_{ac}	is the maximum RMS value of the ac fault current,
X	is equivalent system reactance at the fault point,
R	is equivalent system resistance at the fault point,
t	is time after fault in seconds, and
f	is system frequency of 60 Hz.

The calculation of the symmetrical three phase fault current (I_{ac}) is the result of solving the equivalent impedance model at each fault point. The total asymmetrical current is then calculated separately from Equations (11.5) and (11.6) using I_{ac} and X/R assuming $t = 0.5$ cycle. This provides an accurate estimate of the maximum short circuit current the instant the fault occurs for comparison with the momentary short circuit ratings of a switchgear.

Questions and Problems

11.1 Modeling of electric system loads

The plant voltage under steady-state conditions may normally vary between 90% and 110%. Under transient conditions, the plant voltage may experience wider variations.

Why is it important to know how loads kVA depend on the supply voltage?

11.2 Motor modeling

a) Provide a technical basis for representing an induction motor as a constant kVA load under normal running conditions.

b) Provide a technical basis for representing an induction motor as a constant impedance load under starting conditions.

11.3 Inverter modeling

DC to ac inverter typically used for nuclear plant safety related applications are modeled as constant kVA loads independent of voltage variation. Provide a basis for this approach.

11.4   Voltage drop
Given:
Motor: 3 ph, V = 575 V, I = 0.9 A, pf = 0.6
Source: 575 V
Motor feeder: #12 AWG, 1.9522×10^{-3} Ω / ft, length 31,500 ft, X = 0
Assume base values: V = 575, VA = 1000
   Calculate
a) pu voltage drop
b) pu motor voltage
c) pu voltage regulation.

11.5   Motor starting calculation of accelerating time
   Motor given data:
1000 HP, 1800 rpm, Wk^2 270 lb-ft^2,
Load Wk^2
100% torque = 2916.7 lb-ft^2
   Assuming the value of the accelerating torque to be as shown in Table Q 11.5 below.

Table Q 11.5 Data on Net Accelerating Torque for Motor of Problem 11.5 (Assumed Values Taken from Characteristic Curve, Not Shown)

Period	Speed (%)	T motor (%)	T load (%)	T net (%) (∗)	T net (lb-ft²)
1	0	100	30	—	
				77.5	2260.4
2	25	120	35	—	—
				100	2916.7
3	50	160	45	—	—
				120	2500.0
4	75	190	65	—	—
				62.5	1822.9
5	95	80	80	—	—

(∗) $[T_{motor}\,(n + 1) - T_{load}\,(n + 1)]/2 + (T_{motor}\,(n) - T_{load}\,(n)/2$.

a) Calculate the total motor accelerating time.
b) Is the motor total accelerating time important for the motor protection setting? Explain.
c) Is the motor start time important for the setting of the EDG loading? Explain your answer.

11.6 Short Circuit Calculation
Data:
Transformer 13.8/0.48 V three phase, 1000 kVA, $Z = 6\%$, connected directly to 480 V Switchgear, three phase fault at the bus. Utility source 13.8 kV, 300 MVA, connected to transformer with 13.8 kV cable of $Z = 0.1\ \Omega$.
Load: Equivalent induction motor 1000 kVA @ 480 V, $X_d'' = 25\%$.

References

1 J. L. Blackburn, *Protective Relaying Principles and Applications*, Marcel Decker, Inc., New York.

2 IEEE Std 141, "IEEE Recommended Practice for Electric Power Distribution for Industrial Plants (ANSI)."

3 IEEE Std 142, "IEEE Recommended Practice for Grounding of Industrial and Commercial Power Systems (ANSI)."

4 IEEE Std 242, "IEEE Recommended Practice for Protection and Coordination of Industrial and Commercial Power Systems (ANSI)."

5 IEEE Std 308, "IEEE Standard Criteria for Class 1E Power Systems for Nuclear Power Generating Stations (ANSI)."

6 IEEE 399, "Recommended Practice for Power System Analysis."

7 IEEE Std 765, "IEEE Standard for Preferred Power Supply (PPS) for Nuclear Power Generating Stations (ANSI)."

8 IEEE Std C37.91, "IEEE Guide for Protective Relay Applications to Power Transformers (ANSI)."

9 IEEE Std C37.96, "IEEE Guide for AC Motor Protection."

10 IEEE C37-102, "Guide for AC Generator Protection."

11 IEEE Std C62.92.3, "IEEE Guide for the Application of Neutral Grounding in Electrical Utility Systems, Part III—Generator Auxiliary Systems (ANSI)."

12

Plant Life: Equipment Aging, Life Extension, and Decommissioning

12.1 Nuclear Plant Licensed Life

Most nuclear power plants have operating life times of between 20 and 40 years, and most plants in the United States have been licensed for 40 years. Many plants have requested and obtained a life extension of 20 years additional to the original plant licensed life.

12.2 Importance of Maintenance: The Maintenance Rule (Courtesy of the NRC)

Background

Maintenance is of paramount importance to preserving and extending plant life and to maintaining safe operation. As part of 10 CFR 50 [2], the Nuclear Regulatory Commission (NRC) published 10 CFR 50.65 (commonly referred to as the maintenance rule) on July 10, 1991. The NRC's determination that a maintenance rule was needed arose from the conclusion that proper maintenance is essential to plant safety. As discussed in the Statements of Consideration for this rule, there is a clear link between effective maintenance and safety as it relates to such factors as the number of transients and challenges to safety systems and the associated need for operability, availability, and reliability of safety equipment. In addition, good maintenance is also important in ensuring that failure of other than safety-related structures, systems, and components (SSCs) that could initiate or adversely affect a transient or accident is minimized. Minimizing challenges to safety systems is consistent with the NRC's defense-in-depth.

The objective of 10 CFR 50.65 (referred to hereafter as the maintenance rule or the rule) is to require monitoring of the overall continuing effectiveness of plant operator maintenance programs to ensure that (1) safety-related and

Electrical Systems for Nuclear Power Plants, First Edition. Omar S. Mazzoni.
© 2019 by The Institute of Electrical and Electronic Engineers, Inc. Published 2019 by John Wiley & Sons, Inc.

certain nonsafety-related SSCs are capable of performing their intended functions and (2) for nonsafety-related equipment, failures will not occur that prevent the fulfillment of safety-related functions, and failures resulting in scrams and unnecessary actuations of safety-related systems are minimized. Additional objectives of the maintenance rule are to require that (1) plant operators assess the impact of equipment maintenance on the capability of the plant to perform key plant safety functions and (2) plant operators use the results of the assessment before undertaking maintenance activities at operating nuclear power plants to manage the increase in risk caused by those activities.

Maintenance is also important to ensure that design assumptions and margins in the original design basis are maintained and are not unacceptably degraded. Therefore, nuclear power plant maintenance is important to protecting public health and safety.

The NRC requires in 10 CFR 50.65(a)(1) that power reactor plant operators monitor the performance or condition of SSCs against plant operator-established goals in a manner sufficient to provide reasonable assurance that such SSCs are capable of fulfilling their intended functions. Such goals are to be established commensurate with safety and, where practical, take into account industrywide operating experience. When the performance or condition of an SSC does not meet established goals, appropriate corrective action must be taken.

Nuclear plant operators must monitor the performance or condition of all SSCs associated with storing, controlling, and maintaining spent fuel in a safe condition, in a manner sufficient to provide reasonable assurance that such SSCs are capable of fulfilling their intended functions.

In 10 CFR 50.65(a)(2), the NRC states that monitoring as specified in paragraph (a)(1) is not required when it has been demonstrated that the performance or condition of an SSC is being effectively controlled through the performance of appropriate preventive maintenance, such that the SSC remains capable of performing its intended function.

In 10 CFR 50.65(a)(3), the NRC requires that performance and condition monitoring activities and associated goals and preventive maintenance activities be evaluated at least every refueling cycle provided the interval between evaluations does not exceed 24 months. The evaluations should take into account, where practical, industrywide operating experience. Adjustments should be made where necessary to ensure that the objective of preventing failures of SSCs through maintenance is appropriately balanced against the objective of minimizing unavailability of SSCs due to monitoring or preventive maintenance. Condition monitoring references are given in [13], [23] to [27].

In 10 CFR 50.65(a)(4), the NRC requires that before performing maintenance activities (including but not limited to surveillances, postmaintenance testing, and corrective and preventive maintenance), the plant operator should assess and manage the increase in risk that may result from the proposed maintenance activities. The scope of the assessment may be limited to SSCs that a risk-informed evaluation process has shown to be significant to public health and safety.

In 10 CFR 50.65(b), the NRC states that the scope of the monitoring program specified in 10 CFR 50.65(a)(1) is to include safety-related and nonsafety-related SSCs, as follows:

1. Safety-related SSCs that are relied upon to remain functional during and following design basis events to ensure the integrity of the reactor coolant pressure boundary, the capability to shut down the reactor and maintain it in a safe shutdown condition, or the capability to prevent or mitigate the consequences of accidents that could result in potential offsite exposure.
2. Nonsafety-related structures, systems, or components:
   i) That are relied upon to mitigate accidents or transients or are used in plant emergency operating procedures; or
   ii) Whose failure could prevent safety-related SSCs from fulfilling their safety-related function; or
   iii) Whose failure could cause a reactor scram or actuation of a safety-related system.

The International Atomic Energy Agency (IAEA) has established a series of safety guides and standards constituting a high level of safety for protecting people and the environment. IAEA safety guides present international good practices and increasingly reflects best practices to help users striving to achieve high levels of safety. Pertinent to this regulatory guide, IAEA Safety Guide NS-G-2.6, "Maintenance, Surveillance, and In-service Inspection in Nuclear Power Plants" provides guidance and recommendations on maintenance, surveillance, and in-service inspection activities to ensure that safety-related SSCs are available to perform as designed.

Development of Industry Guideline NUMARC 93-01

In May 1993, the nuclear industry developed NUMARC 93-01, "Industry Guideline for Monitoring the Effectiveness of Maintenance at Nuclear Power Plants" [1], which provides guidance to plant operators on implementation of the maintenance rule. The NRC issued NUREG 1526 [5], to document the evaluation of the new maintenance plan.

The NRC issued the final version of 10 CFR 50.65 by July 1998.

Definition of Maintenance

As discussed in the *Federal Register* notice, "Final Commission Policy Statement on Maintenance at Nuclear Power Plants," dated March 23, 1988 [9], maintenance is defined as the aggregate of those functions required to preserve or restore safety, reliability, and availability of plant SSCs. Maintenance includes not only activities traditionally associated with identifying and correcting actual or potential degraded conditions (i.e., repair, surveillance, diagnostic examination, and preventive measures) but extends to all supporting functions for the conduct of these activities. The activities that form the basis of a maintenance program are also discussed in "Final Commission Policy Statement on Maintenance at Nuclear Power Plants."

Timeliness

NUMARC 93-01 states that activities such as cause determinations and moving SSCs from the 10 CFR 50.65(a)(2) to the (a)(1) category *must be performed in a "timely" manner.*

Plant, System, Train, and Component Monitoring Levels

The extent of monitoring may vary from system to system depending on the system's importance to safety. Some monitoring at the component level may be necessary; however, most of the monitoring can be done at the plant, system, or train level. SSCs with high safety significance, and standby SSCs with low safety significance should be monitored at the system or train level. Normally operating SSCs with low safety significance may be monitored through plant-level performance criteria, including unplanned scrams, safety system actuations, or unplanned capability loss factors. For SSCs monitored in accordance with 10 CFR 50.65(a)(1), additional parameter trending may be necessary to ensure that the problem that caused the SSC to be placed in the 10 CFR 50.65(a)(1) category is being corrected.

Use of Plant Programs

Plant's programs, such as technical specification surveillance testing are normally used to satisfy monitoring requirements.

Use of Reliability-Based Programs

Plants would benefit from the use of reliability-based methods for developing the preventive maintenance programs covered under 10 CFR 50.65(a)(2); see [47] and [48] however, the use of such methods is not required.

Safety Significance Categories

The maintenance rule requires that goals be established commensurate with safety. To implement this requirement, NUMARC 93-01 establishes two safety significance categories, "risk-significant" and "non-risk-significant." Section 9.0 of NUMARC 93-01 describes the process for placing SSCs in either of these two categories. The Statements of Consideration for the rule use the terms "more risk significant" and "less risk significant." NRC Inspection Procedure 71111.12, "Maintenance Effectiveness" [10] uses the terms "high safety significance" and "low safety significance." After discussions with industry representatives, the NRC staff determined that the preferred terminology is "high safety significance" and "low safety significance."

Safety-Significance Ranking Methodology

The NRC staff endorses the use of the SSC safety significance ranking methodology described in NUMARC 93-01 as an acceptable method for meeting the requirements of the maintenance rule (see SECY 95-265 "Response to August 9, 1995, Staff Requirements Memorandum Request to Analyze the Generic Applicability of the Risk Determination Process Used in Implementing the Maintenance Rule," dated November 1, 1995 [11]).

Use of Probabilistic Risk Assessments

NUMARC 93-01 contains multiple references to the use and application of a probabilistic risk assessment (PRA) or a probabilistic safety assessment in a plant's implementation of the maintenance rule. Like other types of engineering analyses used to support the regulatory process, risk analyses must be sound and technically defensible. Sound and technically defensible risk analyses help increase confidence in and the consistency of decision making.

The nuclear plant's maintenance efforts should minimize failures in both safety-related and BOP SSCs that affect safe operation of the plant. The effectiveness of maintenance programs should be maintained for the operational life of the facility.

Maintenance Risk Assessments

The intent of 10 CFR 50.65(a)(4) is to require the plants to conduct assessments before performing maintenance activities on SSCs covered by the maintenance rule and to manage the increase in risk that may result from the proposed activities. RG 1.182 [3], and 64 FR 38551[4] provide guidance for risk in maintenance activities.

12.3   Monitoring Issues Affecting Electrical Equipment, Transformers, Motors, Cable, Control Equipment

Regulatory Guide 1.160 [6], provides general guidelines for complying with 10 CFR 50.65(a)(1). Regulatory Guide 1.218 provides specific guidance for condition monitoring of cables. In particular, this regulatory guide describes a programmatic approach to condition monitoring of electric cable systems and their operating environments, as well as acceptable condition-monitoring techniques. The programmatic approach and condition monitoring may be used to demonstrate compliance with paragraph (a)(1) of the Maintenance Rule. (See [46].)

Cable Failures

Operating experience reveals that the number of cable failures is increasing with plant age and that cable failures are occurring within the plants' 40-year licensing periods. These cable failures have resulted in plant transients and shutdowns, loss of safety functions and redundancy, entries into limiting conditions for operation, and challenges to plant operators. Though in many cases the failed cables were identified through current testing practices, some of the failures may have occurred before the failed condition was identified (i.e., on cables that are not normally energized or tested). Therefore, it is necessary to monitor the condition of electric power cables throughout their installed life through the *use of cable condition-monitoring techniques*. Condition monitoring of cables may be limited to a representative sample of cables, and its frequency may be adjusted based on demonstrated plant-specific cable test results and operating experience.

Improper cable installation (see [10] to [12]) may lead to decrease cable life and eventual cable failures. Cable installation requires monitoring the pulling forces on cables, so as to ensure that the maximum allowable pulling forces are not exceeded. Likewise monitoring of allowable cable sidewall pressure is necessary. Cable sidewall pressure is present when pulling cables around bends.

For cables that are inaccessible or installed underground, appropriate monitoring programs including testing of cables and visual inspection of manholes for water accumulation should be implemented to detect cable system degradation.

Condition monitoring involves the observation, measurement, and trending of one or more condition indicators that can be correlated to the physical condition and/or functional performance of the cable.

12.4   Cable-Monitoring Methods and Techniques (Courtesy of NRC)

(See references [14]–[22], [28]–[36], [38], [39], [40]–[43]) Electric cable condition-monitoring tests may be grouped by whether the inspection or test is performed *in situ* on electric cables in the plant or is a laboratory-type test performed on representative material specimens in a controlled laboratory setting.

These condition-monitoring test techniques may be performed to measure and assess the following:

- *electrical properties* (such as insulation resistance, voltage withstand, dielectric loss/dissipation factor, time domain reflectometry (TDR), and partial discharge),
- *mechanical properties* (such as hardness, elongation at break, and compressive modulus/polymer indenter test),
- *chemical and physical properties* (such as density, oxidation induction time, oxidation induction temperature, and infrared spectroscopy), and
- *physical condition* and appearance.

A combination of condition-monitoring techniques provide significant insights into the condition of cables, as research and experience have shown that no single, nonintrusive, condition-monitoring method currently available, if used alone, is effective to predict the performance of electric cables under accident conditions.

The plant should select condition-monitoring inspection and testing techniques to detect, quantify, and monitor the status of the aging mechanisms that may cause the degradation of the cable system. Consideration should be given to cable insulation and jacket materials, cable construction (e.g., solid/extruded vs. laminated, shielded vs. unshielded), and environmental and operating stressors for each cable system application. By selecting the condition-monitoring techniques that are best suited to the detection and monitoring of the anticipated stressors and associated aging and degradation mechanisms, the plant can more accurately monitor the condition of critical plant cables, assess their operating condition, and implement corrective actions to manage aging and degradation in those cables that are found to be experiencing stressors and aging and degradation rates beyond specified design conditions. The realistic and timely assessment of cable condition is the best means for managing cable degradation and avoiding unexpected early cable failures. Sections 3 and 4.5 of NUREG/CR-7000, "Essential Elements of an Electric Cable Condition Monitoring Program" [6], [37], [44], [45], provide guidance on the selection of condition-monitoring techniques for electric cables. Institute of Electrical and Electronics Engineers (IEEE) Std 1205 " Guide for Assessing, Monitoring, and Mitigating Aging Effects on Class 1E Equipment Used in Nuclear Power Generating Stations" provides guidelines for assessing, monitoring, and mitigating aging degradation effects on Class 1E equipment used in nuclear power generating stations. This IEEE guide also includes informative annexes on aging mechanisms, environmental monitoring, condition monitoring, aging program essential attributes, and example assessments for five types of equipment (including electric cable).

Plant operators may use a number of monitoring techniques to evaluate cable condition. A combination of monitoring techniques may be needed to validate cable performance. Some of the typical condition monitoring techniques and inspection methods that have been or are being used for cable condition monitoring include those described below, which are recommended for use when appropriate. It should be noted that each of the techniques discussed below has advantages and limitations that must be carefully considered when selecting techniques to be used in a condition-monitoring program based on plant-specific cable system design, installation, and operating condition.

1. *Direct Current High-Potential Test (DC High Voltage).* The direct current (DC) high-potential test (HPT) is a pass/fail test applicable to medium-voltage power cables. It is typically used for paper-insulated lead-covered (PILC) cables. Aging mechanisms detected by the HPT comprise thermally induced embrittlement and cracking, radiation-induced embrittlement and cracking, mechanical damage, water treeing, moisture intrusion, and surface contamination.

   Advantages associated with the HPT test are that it does not require access to the entire length of the cable and that the test can potentially detect degradation sites before failure in service. The disadvantages of HPT are that the cable must be disconnected to perform the test and the high voltages used during testing may damage the cable insulation.

   Recent research by the Electric Power Research Institute (EPRI), TR-101245, "Effect of DC Testing on Extruded Cross-Linked Polyethylene Insulated Cables," Volumes 1 and 2 (see references), on medium-voltage cross-linked polyethylene (XLPE) and ethylene propylene rubber (EPR)-insulated cables has shown that a DC HPT of field-aged cables could potentially damage or cause extruded cables, especially field-aged XLPE-insulated cable, to fail prematurely. Among the conclusions reached in the EPRI study are that DC HPTs of field-aged cables can reduce cable life, DC HPTs of field-aged cables generally increase water tree growth, and preenergization DC HPTs of new medium-voltage cable does not significantly reduce cable life. Another disadvantage is that the DC HPT does not provide trendable data.

   Certain cable manufacturers recommend that this test only be done on new cable installations and that it not be performed after the cable has been in service for over 5 years.

2. *Step Voltage Test (DC High Voltage).* The step voltage test (SVT) is a diagnostic test that can be applied to low- and medium-voltage cable. It is typically used for PILC cables. The SVT is capable of detecting aging mechanisms such as thermally induced embrittlement and cracking, radiation-induced embrittlement and cracking, mechanical damage, water treeing, moisture intrusion, and surface contamination.

An advantage of the SVT is that it does not require access to the entire length of the cable. The disadvantages of the SVT are that the cable must be disconnected to perform the test, and the high voltages used during testing may damage the cable insulation. The potential problems with the DC HPT identified by the recent EPRI research study are also applicable to the SVT. Another disadvantage is that the SVT does not provide trendable data.

Certain cable manufacturers recommend that this test only be done on new cable installations and that it not be performed after the cable has been in service for over 5 years.

3. *Very Low Frequency Test.* Very low frequency (VLF) testing methods can be categorized as withstand or diagnostic, which utilize ac voltage signals in the frequency range from 0.01 to 1 Hz. The two most commonly used test voltage signals are the cosine-rectangular and the sinusoidal wave shapes. This test is applicable to shielded medium voltage cables with extruded and laminated dielectric insulation. This test is effective in monitoring the condition of a cable system including the insulation of the terminations and splices if they are included in the test circuit while minimizing or eliminating some potential adverse charging effects of the DC HPT test methods discussed above. The aim of a voltage test is to detect any existing fault, defect, or irregularity prior to a breakdown during service of the cable. Furthermore, the test should not significantly reduce the lifetime of the cable. In withstand testing, the cable must withstand a specified voltage applied across the insulation for a specified period of time without breakdown of the insulation. The magnitude of the withstand voltage is usually greater than that of the applied voltage.

Diagnostic testing is usually performed at lower voltages than withstand tests. This test allows the determination of the relative amount of degradation of a cable system section and establishes whether a cable system section is likely to continue to perform properly in service.

Advantages of the VLF tests are as follows: This test works best when eliminating a few defects from otherwise good cable insulation; dangerous space charges are less likely to be developed in the insulation; and cables may be tested with an ac voltage approximately three times the rated conductor-to-ground voltage with a device comparable in size, weight, and power requirements to a DC test set.

Disadvantages of the VLF tests are as follows: Cables must be disconnected for testing and when testing cables with extensive water tree degradation or partial discharges in the insulation, VLF withstand testing alone may not be conclusive. Additional diagnostic tests that measure the extent of insulation losses will be necessary, and cables must be disconnected for testing.

4. *Illuminated Borescope.* The illuminated borescope (IB) inspection technique is a screening method that can be applied to inaccessible low-voltage

cables, deenergized medium-voltage cables with all types of cable insulation and jacket materials. The IB inspection technique is essentially an optically enhanced visual inspection using the IB tool to visually access cables in otherwise inaccessible conduits and ducts to assess their physical appearance and condition and to identify and locate water intrusion or contamination in the conduits or cable ducts. The IB test is capable of detecting aging mechanisms such as mechanical damage, potential for moisture intrusion, and surface contamination.

Advantages of the IB test are that the test can be relatively easy to implement and can be performed on inaccessible cables to detect the presence of stressors or cable damage and degradation. The disadvantage of the IB test is that it does not provide quantitative data that can be trended. Care should be taken not to damage the cables in conduits when using this inspection technique.

5. *Visual Inspection.* The visual inspection technique for accessible cables is a very simple yet extremely powerful cable condition-monitoring technique for evaluating cable system aging, because physical damage and many degradation mechanisms are readily detectable through sight. Visual inspection can be used to identify changes in physical and visual appearance, surface texture, and damage. Flashlights or magnifiers can aid visual inspection.

The advantages of visual inspection are that it is easy to perform, is minimally intrusive and nondestructive, and can easily detect degradation because of locally adverse conditions. Visual inspection may find surface corona damage of nonshielded medium-voltage cables. The disadvantages are that the cables to be inspected must be visible and accessible; results are not quantitative, making trending very difficult; and appearance is subjective and observations can vary from inspector to inspector.

6. *Compressive Modulus (Polymer Indenter).* This technique is a mechanical properties (hardness) technique that is applicable to polymer jacket and insulation materials, such as polyethylene, EPR, chlorosulfonated polyethylene (CSPE), and neoprene. The compressive modulus technique is most effective at detecting thermally induced embrittlement and radiation-induced embrittlement because it correlates to the phenomenon in elongation at break material test. The technique can detect and monitor the stressor effects of elevated temperature and radiation exposure.

Advantages of the compressive modulus test are that it is relatively easy to perform, it provides trendable data on commonly used cable insulation materials, and results can be correlated to known measures of cable condition. This technique is suitable for assessing short segments of the insulation. The disadvantages are that the cables must be accessible for *in situ* measurements; measurements are made on the outer surface, so the condition of underlying insulation must be inferred; the underlying cable construction, cable geometry, temperature, and humidity affect the results; aging-related changes in the compressive modulus are very small for some polymers until the end of life; the compressive modulus does not give direct correlation to changes in electrical properties (such as insulation resistance and dielectric strength); and the test has limited usefulness for XLPE cables. However, it can be used on XLPE cable with neoprene or CSPE jackets to provide a leading indication of damage.

7. *Dielectric Loss Dissipation Factor (Power Factor).* The dielectric loss-dissipation factor or power factor test (tan δ test) can be used to diagnose problems in medium-voltage cables. The dielectric loss-dissipation factor test has the ability to detect thermally induced cracking, radiation-induced cracking, mechanical damage, water treeing, moisture intrusion, and surface contamination.

   Advantages associated with the dielectric loss-dissipation factor technique are that it provides trendable data on commonly used cable insulation materials, it does not require access to the entire cable, and the results can be correlated to known measures of cable condition. Disadvantages include that the end terminations of the cable must be disconnected to perform the test, the test is only applicable to cables that have shielded or sheath construction because it requires a defined ground return path of the loss (leakage) current back to the test set (supply source), the test should not be performed on low-voltage and medium voltage unshielded cables because of physical safety concerns and unreliable test results resulting from an undefined ground return path, and the amount of capacitance in the cable circuit limits the test such that standard test equipment cannot test very long and larger conductor cables.

8. *Insulation Resistance.* The insulation resistance test is a diagnostic test that is relatively effective with low-voltage cables using all types of insulation and jacket materials. The insulation resistance test is a standard test used to measure the dielectric integrity of cable insulation. Because of its sensitivity to temperature and humidity, it frequently is used as a pass/fail test because of the difficulties in obtaining an accurate and consistent absolute insulation resistance measurement.

   Advantages of the test are that it does not require access to the entire cable, it does not need to be corrected for temperature effects, and it can provide trendable data. The disadvantages are that the end terminations of the cable must be disconnected to perform the test, the test is not as sensitive to insulation degradation as other electrical properties techniques, and leakage currents are very small and sensitive to surrounding environmental conditions, making it difficult to measure accurately.

9. *Partial Discharge Test.* The partial discharge test (PDT) is a diagnostic test that applies to medium-voltage shielded cables using all types of cable

insulation and jacket materials. Aging mechanisms detected by the PDT provide information that can be used to detect the presence of thermally induced cracking, mechanical damage, and radiation-induced cracking. Partial discharge is a useful tool for concentric neutral cables for determining whether splice degradation has occurred. Partial discharge may not be effective for tape-shielded cable systems because of attenuation of the signal from shield corrosion.

Advantages of the PDT are that it does not require access to the entire length of the cable, it identifies the significant partial discharge sites in an insulation system, it provides information on the severity of the insulation defects, and it gives information on the location of each of the significant partial discharge sites (and insulation defects). Disadvantages are that the end terminations of the cable must be disconnected to perform the test, performance of the PDT is complex and requires a high skill level, the interpretation of PDT results requires a very high skill level and training, and the high testing voltage applied during the PDT has the potential to weaken and permanently damage the cable insulation. Also, because nearby operating electrical equipment in a plant environment could cause noise interference with the test, this test is most successful on shielded cables.

10. *Time Domain Reflectometry.* The TDR test is a diagnostic test that can be implemented on low- and medium-voltage cables using all types of cable insulation and jacket materials. The TDR test is able to detect the occurrence of aging degradation such as thermally induced cracking, radiation-induced cracking, and severe mechanical damage that have an effect on cable impedance. The TDR test can also identify the presence of water and its location along a cable run, the location and severity of electrical faults, and the location and severity of insulation damage. TDR testing may not be effective for tape-shielded cable systems because of attenuation of the signal from shield corrosion.

Advantages of the TDR test are that it provides useful information for identifying and locating potential defects and discontinuities in a cable that may indicate severe insulation degradation or impending cable fault, it is nondestructive, it can be performed *in situ* from one end of a cable, and data can be trended against a baseline reflectogram. Disadvantages are that the end terminations of the cable must be disconnected to perform the test; training and experience are required for best results; and transient conditions, such as immersion, are only detectable when present during the TDR test.

11. *Frequency Domain Reflectometry.* Frequency domain reflectometry (FDR) is a diagnostic test based on transmission line theory. An example is the LIRA method (line resonance analysis), which is based on FDR techniques. It is applicable to low- and medium-voltage cables of all types of cable

insulation and jacket materials. The FDR test can detect aging mechanisms such as thermally induced embrittlement and cracking, radiation-induced embrittlement and cracking, and severe mechanical damage to the cable insulation.

Advantages of the FDR test are that it can be performed *in situ* without disconnecting the cable, the test requires only a single access point, the analysis of results can account for the effects of loads attached to the cable, and it can accurately identify the site of localized degradation. Disadvantages are that the test is not simple to perform or interpret, and training and experience are needed to obtain meaningful results. It is relatively new technology.

12. *Infrared Imaging Thermography.* The infrared imaging thermography technique is a nondestructive, noncontact, electronically enhanced visual inspection technique for electrical equipment that is simple to perform and valuable in identifying potentially damaging service conditions where elevated temperatures are present. This technique is applicable to all cables. The infrared imaging test is able to provide information useful for detecting aging degradation such as thermally induced embrittlement and cracking. Infrared imaging provides a useful tool for identifying temperature hotspots that could lead to accelerated degradation of electric cable systems or that indicate high-resistance electrical joints in electrical connectors and splices because of loosening, dirt or contamination, or corrosion. The instrument's high-resolution temperature detection capabilities combined with image storage and analysis software make it possible to trend the thermal data obtained.

Advantages of the infrared imaging thermography technique are that it is relatively easy to perform, properly corrected data can identify the temperatures and location of hotspots, measurements can be made when the circuit is operating with a full load, data may be stored and trended with appropriate software, the test is nondestructive and nonintrusive, and it does not require the cable system under test to be disconnected. Disadvantages are that it requires training and experience for best results, measurements made when the circuit is operating at load can lead to physical safety concerns, high-end imagers and analysis software are expensive, and the cables and accessories to be monitored must be visually accessible.

12.5 Further Information on Cable Testing

Further information describing the selection and performance of many different types of cable condition-monitoring techniques, including *in situ* methods and laboratory tests, appears in references [8], [9], [14]–[22], [28]–[36], [38], [39], [40]–[43].

12.6   Switchyard Maintenance Activities

Plant management should have the ability to control and monitor all maintenance activities performed in the high-voltage switchyard.

12.7   Emergency Diesel Generators

Industry-sponsored PRAs have shown the safety significance of emergency ac power sources. The station blackout rule (10 CFR 50.63, "Loss of All Alternating Current Power") requires plant-specific coping analyses to ensure that a plant can withstand a total loss of ac power for a specified duration and to determine appropriate actions to mitigate the effects of a total loss of ac power. During the station blackout reviews, most plant operators (1) committed to implementing an emergency diesel generator reliability program in accordance with NRC regulatory guidance but reserved the option to later adopt the outcome of Generic Issue B-56 resolution, and (2) stated that they had an equivalent program or will implement one. Subsequently, utilities docketed commitments to maintain their selected target reliability values (i.e., maintain the emergency diesel generator target reliability of 0.95 or 0.975). Those values could be used as a goal or as a performance criterion for emergency diesel generator reliability under the maintenance rule.

Emergency diesel generator unavailability values were also assumed in plant-specific individual plant examination analyses. These values should be compared to the plant-specific emergency diesel generator unavailability data regularly monitored and reported as industrywide plant performance information. These values could also be used as the basis for a goal or performance criterion under the maintenance rule. In addition, in accordance with 10 CFR 50.65(a)(3), plant operators must periodically balance the unavailability and reliability of the emergency diesel generators.

12.8   Interpretation of "Standby"

In NUMARC 93-01, standby SSCs of low safety significance must have SSC-specific performance criteria or goals, similar to SSCs of high safety significance. NUMARC 93-01 provides a definition of standby. Some plant operators have improperly interpreted this definition to mean that SSCs that are energized are normally operating. As stated in NUMARC 93-01, *if the SSC performs its intended function only when initiated by either an automatic or manual demand signal, the SSC is in standby.*

Normally operating SSCs are those whose failure would be readily apparent (e.g., a pump failure results in loss of flow that causes a trip). Standby SSCs are those whose failure would not become apparent until the next demand, actuation, or surveillance. Only those SSCs of low safety significance whose failure would be readily apparent (because they are normally operating) should be monitored by plant-level criteria.

SSCs may have both normally operating and standby functions. To adequately monitor the effectiveness of maintenance for the SSCs associated with standby functions, plant operators should develop SSC-specific performance criteria or goals, or condition monitoring.

12.9 Normally Operating SSCs of Low Safety Significance

Cause Determinations

For all SSCs that are being monitored using plant-level performance criteria (i.e., normally operating SSCs of low safety significance), the NRC staff's position is that a cause determination should be performed whenever any of these performance criteria are exceeded (i.e., failed) to determine which SSC caused the criterion to be exceeded or whether the failure was a repetitive maintenance preventable functional failure (MPFF). As part of the cause determination, it would also be necessary to determine whether the SSC was within the scope of the maintenance rule and, if so, whether corrective action and monitoring (tracking, trending, goal setting) under 10 CFR 50.65(a)(1) should be performed.

12.10 Establishing SSC-Specific Performance Criteria

The maintenance rule requires that plant operators monitor the effectiveness of maintenance for all SSCs within the scope of the rule. NUMARC 93-01 allows plant operators to monitor SSCs of low safety significance with plant-level criteria. NUMARC 93-01 notes that some normally operating SSCs of low safety significance cannot be practically monitored by plant-level criteria. Plant operators should ensure that the plant-level criteria established do effectively monitor the maintenance performance of the normally operating SSCs of low safety significance, or they should establish SSC-specific performance criteria or goals or use condition monitoring.

For example, a plant operator determined that the rod position indication system and the spent fuel pool pit cooling system were within the scope of the maintenance rule because they were safety related at the plant operator's site. None of the three plant-level performance criteria described in NUMARC 93-01 (unplanned scrams, unplanned capability loss factor, or unplanned safety system actuations) would monitor the effectiveness of maintenance on these systems. Therefore, plant operators should establish additional plant-level performance criteria or system-specific performance criteria.

12.11 Clarification of MPFFs Related to Design Deficiencies

The third paragraph of Section 9.4.5 of NUMARC 93-01 provides guidance on the plant operator's options following a failure and on whether, as a result of the plant operator's corrective actions, subsequent failures would be considered MPFFs. In particular, this paragraph addresses failures caused by design deficiencies. Ideally, plant operators would modify the design to eliminate the poorly designed equipment. However, if the plant operator determines that such an approach is not cost effective (e.g., the cost of modification is prohibitive), the plant operator has two options:

1. Replace or repair the failed equipment and adjust the preventive maintenance program as necessary to prevent recurrence of the failure. Subsequent failures of the same type that are caused by inadequate corrective or preventive maintenance would be MPFFs and could be repetitive MPFFs.
2. Perform an evaluation that demonstrates that the equipment can be run to failure (as described in Section 9.3.3 of NUMARC 93-01). If the equipment can be run to failure, the plant operator may replace or repair the failed equipment, but adjustments to the preventive maintenance program are not necessary and subsequent failures would not be MPFFs.

12.12 Scope of the Hazards to be Considered during Power Operations

NUMARC 93-01 provides guidance to plant operators on the scope of hazard groups to be considered for the 10 CFR 50.65(a)(4) assessment provision during power-operating conditions. Section 11.3.3 of NUMARC 93-01 specifically considers internal events, internal floods, and internal fires for assessment. Section 11.3.4.2 of NUMARC 93-01 also considers weather, external flooding, and other external impacts if such conditions are imminent or have a high probability of occurring during the planned out-of-service duration. The NRC staff considers these two sections of NUMARC 93-01 to encompass the scope of hazards that plant operators should consider during power operation to perform an adequate assessment of the potential impact of risk that may result from proposed maintenance activities.

12.13 Scope of Initiators to be Considered for Shutdown Conditions

NUMARC 93-01 provides guidance to plant operators on the scope of hazard groups to be considered for the 10 CFR 50.65(a)(4) assessment provision during shutdown conditions. Section 11.3.6 of NUMARC 93-01 specifically considers internal events for assessment as well as weather, external flooding, and other external impacts if such conditions are imminent or have a high probability of occurring during the planned out-of-service duration. The NRC staff considers this section of NUMARC 93-01 to encompass the scope of hazards that plant operators should consider during shutdown conditions to perform an adequate assessment of the potential impact of risk that may result from proposed maintenance activities.

12.14 Fire Scenario Success Path(s)

The last paragraph of Section 11.3.3.1 of NUMARC 93-01 states that some fire scenarios have no success paths available. The NRC does not agree with this statement within its context in NUMARC 93-01. Each plant is required by 10 CFR 50.48, "Fire Protection" to identify one train of safe-shutdown capability free of fire damage, such that the plant can be safely shut down in the event of a fire. When maintenance activities are conducted on the protected train, the staff's position is that plant operators should follow the guidance in Section 11.3.4.3 of NUMARC 93-01.

12.15 Establishing Action Thresholds Based on Quantitative Considerations

In Section 11.3.7.2 of NUMARC 93-01, the authors suggest the value "10^{-3}/year" as a ceiling for configuration-specific core damage frequency. At this time, the NRC neither endorses nor disapproves of the 10^{-3}/year value.

12.16 SSCs Considered under 10 CFR 50.65(a)(1)

In 10 CFR 50.65(a)(1), the NRC requires that goal setting and monitoring be established for all SSCs within the scope of the rule except for those SSCs whose performance or condition is adequately controlled through the performance of appropriate preventive maintenance as described in 10 CFR 50.65(a)(2). NUMARC 93-01 initially places all SSCs under 10 CFR 50.65(a)(2) and only moves them to consideration under 10 CFR 50.65(a)(1) if experience indicates that the performance or condition is not adequately controlled through preventive maintenance, as evidenced by the failure to meet a performance criterion or by experiencing a repetitive MPFF. Therefore, the 10 CFR 50.65(a)(1) category could be used as a tool to focus attention on those SSCs that need to be monitored more closely. It is possible that no (or very few) SSCs would be handled under the requirements of 10 CFR 50.65(a)(1). However, the rule does not require this approach. Plant operators could also take the approach that all (or most) SSCs would be handled under 10 CFR 50.65(a)(1), and none (or very few) would be considered under 10 CFR 50.65(a)(2). Plant operators may take either approach.

The number of SSCs monitored under 10 CFR 50.65(a)(1) can vary greatly because of factors that have nothing to do with the quality of the plant operator's maintenance activities. For example, two identical plants with equally effective maintenance programs could have different numbers of SSCs monitored under 10 CFR 50.65(a)(1) because of differences in the way system boundaries are defined (e.g., a system with three trains may be defined as one system at one plant while the same system may be defined as three separate systems at an identical plant) or because of differences in the way performance criteria are defined at the two plants (e.g., a plant operator that takes a very conservative approach to monitoring against the performance criteria would have more SSCs in the 10 CFR 50.65(a)(1) category). If there is some doubt about whether a particular SSC should be monitored under 10 CFR 50.65(a)(1) or (a)(2), the conservative approach would be to monitor the SSC under 10 CFR 50.65(a)(1).

12.17 Inclusion of Electrical Distribution Equipment

The monitoring efforts under the maintenance rule, as defined in 10 CFR 50.65(b), encompass those SSCs that directly and significantly affect plant operations, regardless of which organization actually performs the maintenance activities. Maintenance activities that occur in the switchyard can directly affect plant operations; as a result, electrical distribution equipment out to the first intertie with the offsite distribution system (i.e., equipment in the switchyard) should be considered for inclusion as defined in 10 CFR 50.65(b).

12.18 The License Renewal Rule

The NRC requirements for plant life extension are contained in to Title 10, Part 54, of the Code of Federal Regulations (10 CFR Part 54) [7], "Requirements for Renewal of Operating Licenses for Nuclear Power Plants." These requirements specify how power reactor plant operators are to demonstrate that they are managing the effects of aging. As such, IP-71003 is primarily a part of the programs conducted under NRC Inspection Manual Chapter (IMC) 2516, "Policy and Guidance for the License Renewal Inspection Programs," and the related inspections are also referenced as an "infrequent activity" in Appendix C to IMC 2515, "Light-Water Reactor Inspection Program—Operations Phase."

There is a clear safety objective related to monitoring passive components and their intended safety-related functions within the scope of the License Renewal Rule [10 CFR 54.4(a)(1)].

In addition, 10 CFR 54.4(a)(2) includes the requirement to monitor nonsafety-related components that have the potential to affect safety-related equipment. Other components within scope of the License Renewal Rule also have a clear regulatory basis related to the intended function, such as station blackout power [10 CFR 54.4(a)(3)]. For example, the failure of passive components could be a common-mode failure for certain safety systems.

Other components within scope of the License Renewal Rule also have a clear regulatory basis related to the intended function, such as station blackout power [10 CFR 54.4(a)(3)]. For example, the failure of passive components could be a common-mode failure for certain safety systems.

12.19 Interpretation of Aging

Aging is defined as a continuing time-dependent degradation of material due to service conditions, including normal operation and transient conditions. It is common experience that over long periods of time, there is a gradual change in the properties of materials. These changes can affect the capability of engineered components, systems, or structures to perform their required function. Not all changes are deleterious, but it is commonly observed that aging processes normally involve a gradual reduction in performance capability.

All components in a nuclear power plant can suffer from aging and can partially or totally lose their designed function. Aging is not only of concern for active components (for which the probability of malfunction increases with time) but also for passive ones, since the safety margin is being reduced toward the lowest allowable level.

12.20 Effects of Plant Aging

The main aging effects of concern are changes in physical properties (e.g., electric conductivity), irradiation embrittlement, thermal embrittlement, creep, fatigue, corrosion (including erosion and cracking assisted by corrosion), and wear (e.g., fretting and cracking assisted by wear, such as fretting fatigue).

The term "aging" thus represents the cumulative changes over time that may occur within a component or structure because of one or more of these factors. From this perspective, it is clear that this is a complex process that begins as soon as a component or structure leaves its production facilities, and that it continues throughout its service life.

Questions and Problems

12.1 Decreased insulation life, caused class 1E power cable feeding a safety-related pump motor to develop a failure that in turn caused the safety pump to shut down due to fast acting operation of the cable differential relaying. Preliminary investigation revealed the insulation had failed. The cable had not been tested since initial start of operation, 20 years ago. The pant replaced the failed cable and continued operations. To keep the shutdown time to a minimum, the plant was restarted without testing of the redundant cables.

- Was the plant approach adequate in terms of the plan for resuming operations?
- Do you think the cable failure was age related?
- What were the significant safety-related aspects involved in the plant's operating actions following the cable failure?
- What should the plant have done in your opinion?

12.2 A nuclear plant's high voltage switchyard was installed 30 years ago, it is located by the sea and never had its grounding grid tested after the first installation test.

Prompted by concerns about switchyard conductor's corrosion and reduced life, and in accordance with IEEE 80 requirements, an outside consultant strongly recommended the plant to perform testing at the earliest opportunity. As the plant was interested in saving the funds necessary for the test, the consultant was asked to provide the consequences of not testing at all.

- What might have been the consultant's response?

12.3 A nuclear plant power transformer is only required to operate upon a station blackout condition. As such, the plant has designed the transformer to be normally deenergized, but ready to be energized. In accordance with the plant surveillance procedures, the transformer is periodically tested.

An outside consultant has recommended the plant to institute a change to have the transformer primary winding continuously energized.

- What might be the reasons for the consultant's recommendation?

12.4 A nuclear plant had a safety-related underground cable duct running between two manholes. The duct was designed with slope down from each manhole, such that at an intermediate point between the manholes there was a low point.

- Was the underground cable duct installation adequate in terms of facilitating the cable installation?

- Was the intermediate low point a concern that might result in effects on cable life?

12.5 A nuclear plant had a safety-related, directly buried, 3 mile long underground cable and wanted to institute a test program that would ensure adequate monitoring of the cable insulation condition. Obviously due to the underground portion of the cable being inaccessible, the testing program could only consider access to the cable at the ends, where the cable would be exposed.
- Provide a proposal for an effective cable insulation life monitoring.

12.6 A nuclear plant underground cable installation was performed by a contractor who failed to record the cable-pulling forces for some of the class 1E cable installed in underground ducts. An outside consultant was concerned in that the absence of proper records did not allow for the verification that pulling forces were below allowable maximum levels and that the insulation life may have been detrimentally affected.
- Was the consultant right? If so, why?
- What tests could be recommended to assess the condition of the installed cables?

References

1 NUMARC 93-01, "Industry Guideline for Monitoring the Effectiveness of Maintenance at Nuclear Power Plants," Revision 4A, Nuclear Energy Institute, Washington, DC, April 2011.
2 10 CFR Part 50, "Domestic Licensing of Production and Utilization Facilities," US Nuclear Regulatory Commission, Washington, DC.
3 Regulatory Guide, 1.182, "Assessing and Managing Risk before Maintenance Activities at Nuclear Power Plants," US Nuclear Regulatory Commission, Washington, DC.
4 64 FR 38551, "Monitoring the Effectiveness of Maintenance at Nuclear Power Plants," *Federal Register*, vol. 64, no. 137, p. 38551, July 19, 1999.
5 NUREG-1526, "Lessons Learned from Early Implementation of the Maintenance Rule at Nine Nuclear Power Plants," US Nuclear Regulatory Commission, Washington, DC, June 1995.
6 Regulatory Guide 1.160, "Monitoring the Effectiveness of Maintenance at Nuclear Power Plants."
7 10 CFR Part 54, Requirements for Renewal of Operating Licenses for Nuclear Power Plants.
8 M. Celina, K. T. Gillen, and R. L. Clough, "Inverse Temperature and Annealing Phenomena during Degradation of Crosslinked Polyolefins," *Polymer Degradation and Stability*, vol. 61, no. 2, pp. 231–244, 1998.

9 M. Celina, K. T. Gillen, J. Wise, and R. L. Clough, "Anomalous Aging Phenomena in a Crosslinked Polyolefin Cable Insulation," *Radiation Physics and Chemistry*, vol. 48, no. 5 pp. 613–626, November 1996.

10 NRC Information Notice 2002-12, "Submerged Electrical Cables," US Nuclear Regulatory Commission, Washington, DC, March 21, 2002.

11 NRC Information Notice 1989-63, "Possible Submergence of Electrical Circuits Located Above the Flood Level Because of Water Intrusion and Lack of Drainage," US Nuclear Regulatory Commission, Washington, DC, September 5, 1989.

12 NRC Information Notice 2010-26, "Submerged Electrical Cables," US Nuclear Regulatory Commission, Washington, DC, December 2, 2010.

13 Generic Letter 2007-01, "Inaccessible or Underground Power Cable Failures that Disable Accident Mitigation Systems or Cause Plant Transients," US Nuclear Regulatory Commission, Washington DC, February 7, 2007.

14 NUREG/CR-7000, "Essential Elements of an Electric Cable Condition Monitoring Program," US Nuclear Regulatory Commission Washington, DC, January 2010.

15 IEEE Std 1205, "IEEE Guide for Assessing, Monitoring, and Mitigating Aging Effects on Class 1E Equipment Used in Nuclear Power Generating Stations," Institute of Electrical and Electronics Engineers, New York, NY.

16 TR-101245, "Effect of DC Testing on Extruded Cross-Linked Polyethylene Insulated Cables," Volumes 1 (1993) and 2 (1995). Electric Power research Institute, Palo Alto, CA.

17 NUREG/CR-6704, "Assessment of Environmental Qualification Practices and Condition Monitoring Techniques for Low-Voltage Electric Cables," Volume 2: "Condition Monitoring Test Results," US Nuclear Regulatory Commission, Washington, DC, February 2001.

18 SAND96-0344, "Aging Management Guideline for Commercial Nuclear Power Plants—Electrical Cable and Terminations," Sandia National Laboratories, Albuquerque, NM.

19 IEEE Std 400, "IEEE Guide for Field Testing and Evaluation of the Insulation of Shielded Power Cable Systems" Institute of Electrical and Electronics Engineers, New York, NY.

20 IEEE Std 400.1, "IEEE Guide for Field Testing of Laminated Dielectric, Shielded Power Cable Systems Rated 5 kV and Above with High Direct Current Voltage" Institute of Electrical and Electronics Engineers, New York, NY.

21 IEEE Std 400.2, "IEEE Guide for Field Testing of Shielded Power Cable Systems Using Very Low Frequency (VLF)," Institute of Electrical and Electronics Engineers, New York, NY.

22 IEEE Std 400.3, "IEEE Guide for Partial Discharge Testing of Shielded Power Cable Systems in a Field Environment," Institute of Electrical and Electronics Engineers, New York, NY

23 NUREG/CR-5772, "Aging, Condition Monitoring, and Loss-of-Coolant Accident (LOCA) Tests of Class 1E Electrical Cables," November 1992.

24 NUREG/CR-6794, "Evaluation of Aging and Environmental Qualification Practices for Power Cables Used in Nuclear Power Plants," January 2003.

25 NUREG/CR-6788, "Evaluation of Aging and Qualification Practices for Cable Splices Used in Nuclear Power Plants," September 2002.

26 G. Toman, and J. B. Gardner, "Development of a Non-Destructive Cable Insulation Test," in NUREG/CP-0135, "Workshop on Environmental Qualification of Electric Equipment," November 1993.

27 G. Toman, "Oxidation Induction Time Concepts," in NUREG/CP-0135, "Workshop on Environmental Qualification of Electric Equipment," November 1993.

28 ASTM D257, "Standard Test Methods for DC Resistance or Conductance of Insulating Materials."

29 ASTM D150, "Standard Test Methods for AC Loss Characteristics and Permittivity (Dielectric Constant) of Solid Electrical Insulation."

30 ASTM D470, "Standard Test Methods for Crosslinked Insulations and Jackets for Wire and Cable."

31 ASTM D2633, "Standard Test Methods for Thermoplastic Insulations and Jackets for Wire and Cable."

32 ASTM D638, "Standard Test Method for Tensile Properties of Plastics."

33 ASTM D412, "Standard Test Methods for Vulcanized Rubber and Thermoplastic Rubbers and Thermoplastic Elastomers—Tension."

34 ASTM D3895, "Test Method for Oxidative-Induction Time of Polyolefins by Digital Scanning Calorimetry."

35 ASTM D792, "Standard Test Methods for Density and Specific Gravity (Relative Density) of Plastics by Displacement."

36 ASTM D1505, "Standard Test Method for Density of Plastics by the Density-Gradient Technique."

37 EPRI TR-109619, "Guideline for the Management of Adverse Localized Equipment Environments."

38 EPRI TR-103834-P1-2, "Effects of Moisture on the Life of Power Plant Cables."

39 IAEA TECDOC 1188, "Assessment and Management of Aging of Major Nuclear Power Plant Components Important to Safety: In-Containment Instrumentation and Control Cables," Volumes 1 and 2.

40 P. F. Fantoni, "Wire System Aging Condition Monitoring and Fault Detection Using Line Resonance Analysis," presentation to the NRC, Organization for Economic Co-operation and Development Halden Reactor Project, Institute for Energiteknikk, Halden, Norway, April 5, 2005.

41 K. T. Gillen, M. Celina, and R. L. Clough, "Density Measurements as a Condition Monitoring Approach for Following the Aging of Nuclear Power Plant Cable Materials," *Radiation Physics and Chemistry*, vol. 56, no. 4, pp. 429–447, October 1999.

42 J. P. Steiner, and F. D. Martzloff, "Partial Discharges in Low-Voltage Cables," paper presented at the IEEE International Symposium on Electrical Insulation, Toronto, Canada, IEEE, New York, NY. June 1990.

43 Y. Tian, P. L. Lewin, A. E. Davies, and Z. Richardson "Acoustic Emission Detection of Partial Discharges in Polymeric Insulation," Paper presented at 11th International Symposium on High Voltage Engineering, IEEE, New York, NY. September 2000.

44 ASME NQA-1, "Quality Assurance Requirements for Nuclear Facilities Applications."

45 IEEE Std 323, "IEEE Standard for Qualifying Class 1E Equipment for Nuclear Power Generating Stations."

46 IEEE Std 338, "IEEE Standard Criteria for the Periodic Surveillance Testing of Nuclear Power Generating Station Safety Systems."

47 IEEE Std 352, "IEEE Guide for General Principles of Reliability Analysis of Nuclear Power Generating Station Safety Systems."

48 IEEE Std 577, "IEEE Standard Requirements for Reliability Analysis in the Design and Operation of Safety Systems for Nuclear Power Generating Stations."

13

Electrical and Control Systems Inspections

13.1 Purpose of Inspections

This chapter reviews the general requirements for inspecting a nuclear plant electrical systems in accordance with established industry standards and the regulations and approach of the US Nuclear Regulatory Commission (NRC). The Standard Review Plan [1] is utilized as the prime document for conducting the inspection.

13.2 Objectives of Inspections

Performance capability of a nuclear power plant's electrical system is assessed by reviewing the following:

- initial basic design performance capabilities of the electrical system,
- current as built design performance capabilities of the electrical system,
- adequacy of the existing electrical system in meeting the initial and current design performance capabilities, and
- engineering organization performance by the licensee is also reviewed to determine adequacy in the following areas:
  - procurement,
  - modifications, and
  - operations.

Additionally, the inspectors will have assessed the adequacy of the engineering organization support for the operating plant.

Electrical Systems for Nuclear Power Plants, First Edition. Omar S. Mazzoni.
© 2019 by The Institute of Electrical and Electronic Engineers, Inc. Published 2019 by John Wiley & Sons, Inc.

13.3 Areas of Review

The specific areas of review are as follows:

a) The plant's description of the off-site power system with regard to the inter-relationships between the nuclear unit, the utility grid, and the interconnecting grids.
b) The plant's description of the on-site power systems with regard to the availability of sufficient power to mitigate design-basis events given a loss of the off-site power system and a single failure in the on-site power system.
c) The plant's description of the capability to withstand and recover from a station blackout event of a specified duration.
d) The acceptance criteria to be implemented in the design of the above systems.

Aspects requiring attention:

a) overloaded Class 1E power supplies,
b) inadequate system voltage regulation,
c) inadequate system protection and coordination,
d) modifications installed without consideration of the design basis of safety system requirements and functionality,
e) temporary modifications with unreviewed safety questions,
f) inadequate root cause analysis,
g) inadequate corrective action, and
h) inadequate testing and surveillance.

13.4 Typical Approach to the Review

Typically, the NRC performs a probability analysis to determine the set of components to be inspected. Components exhibiting a relatively high level of risk are selected.

Utilizing the system one-line diagram, safety analysis report (SAR), system descriptions, and procedures, the inspectors review the following electrical equipment characteristics and attributes:

1. the design bases, criteria, standards, regulatory guides, and technical positions implemented in the design of the electric power systems, including the extent to which these criteria and guidelines were followed and the extent of conformance of the design to each requirement,
2. basic performance capabilities of the electrical distribution system,
3. typical system lineups for normal and abnormal, and emergency operation,
4. concept of preferred and standby/alternate sources,
5. bus transfers and load shedding/sequencing, and
6. electrical equipment attributes including

- the critical characteristics, including current, voltage, power, power factor, voltage regulation, protective device coordination,
- the testing and surveillance,
- typical maintenance requirements,
- licensee procedures, standards, regulations, and manufacturers' guidance and requirements,
- short circuit study methodology,
- voltage regulation study and load study,
- electrical equipment set point calculation methodologies and accuracy,
- circuit breaker, protective relay and fuse coordination characteristics,
- spurious signal protection methodologies,
- the separation requirements (common enclosure concerns) for electrical equipment and cables,
- electrical equipment seismic/environmental qualification requirements,
- fire protection,
- emergency diesel generator service water, fuel/lube oil transfer and storage, starting air system, heating, ventilation, and air conditioning
- adequacy of environmental conditions,
- adequacy of protection against floods and water intrusion,
- adequacy of seismic protection,
- interface with the transmission grid, including voltage, frequency, and system stability, and
- adequacy of station blackout design and requirements.

Utilizing the system one-line diagram, SAR, system descriptions and procedures, the inspectors review the following electrical engineering organization issues:

- adequacy of procurement,
- adequacy of modifications,
- adequacy of root cause analysis,
- adequacy of design control,
- adequacy of corrective action,
- adequacy of evaluation of industry operating experience,
- adequacy of design organization and the interface with operations and regulatory groups in the plant,
- adequacy of training in electrical engineering,
- adequacy of electrical equipment maintenance, and
- adequacy of coordination with the transmission grid operators.

13.5 Acceptance Criteria

Inspection results should be evaluated against the following acceptance criteria:

Criteria	Title	Remarks
General Design Criteria (GDC) Appendix A to 10 CFR 50		
a. GDC 2	Design Bases for Protection Against Natural Phenomena	
b. GDC 4	Environmental and Dynamic Effects Design Bases	
c. GDC 5	Sharing of Structures, Systems, and components	
d. GDC 17	Electric Power Systems	
e. GDC 18	Inspection and Testing of Electric Power Systems	
f. GDC 33	Reactor Coolant Make up	As they relate to the operation of electric power systems, encompassed in GDC 17 to ensure the safety function of the systems described
g. GDC 34	Residual Heat Removal	
h. GDC 35	Emergency Core Cooling	
i. GDC 38	Containment Heat Removal	
i. GDC 41	Containment Heat Removal	
i. GDC 44	Containment Heat Removal	
f. GDC 33	Reactor Coolant Makeup	
g. GDC 34	Residual Heat Removal	
h. GDC 35	Emergency Core Cooling	
i. GDC 38	Containment Heat Removal	
j. GDC 41	Containment Atmosphere	
k. GDC 44	Cleanup Cooling Water	
l. GDC 50	Containment Design Bases	

Questions and Problems

13.1 Inspections are normally conducted by the US NRC and by the plant operators. What is the term used for the inspections conducted by the plant management?

13.2 Referring to the NRC inspection plan, what method does the NRC utilize for selecting the set of components to be inspected?

13.3 Referring to the NRC inspection plan, are components selected always Class 1E?

13.4   What is the NRC basis for selecting the systems to be inspected?

13.5   An NRC team concentrated on equipment Division 1, out of two redundant divisions. If several issues were found during the inspection related to division 1, what should the inspectors recommend regarding expanding the scope of the inspection?

13.6   An inspector witnessed a battery discharge test and observed that the plant technician performed preconditioning (maintenance previous to the test, such as cleaning of terminals and adding water).
- Should the preconditioning be allowed?
- What documentation should the inspector review to ascertain this aspect of the test?

Reference

1 US NRC Standard Review Plan, NUREG-0800, Section 8, "Electric Power."

Appendix 1

Abbreviations

Sources of Information: US NRC NUREG-0544, NRC Collection of Abbreviations; IEEE Standard 100, The Authoritative Dictionary of IEEE Standards Terms; IEC Electropiedia, "IEC 60050 – International Electrotechnical Vocabulary" (www.electropedia.org).

AA	access authorization
Aac	alternate ac
AAM	airborne activity monitor
AB	auxiliary boiler
ABT	automatic bus transfer
ABWR	advanced boiling water reactor
ac	alternating current
ACB	air-operated circuit breaker
ACF	ampacity correction factor
	automatic control feature,
ACP	auxiliary control panel
A/D	analog to digital
ADAMS	Agencywide Documents Access and Management System (an NRC abbreviation)
ADAS	automatic data acquisition and storage
ADC	apparent discrepancy notification
ADP	automated data processing
AEC	Atomic Energy Commission, US (became ERDA and NRC)
AEOF	alternate emergency operations facility
AF	auxiliary feed
AF/AL	as found, as left
AFC	automatic flow control
AGVC	automatic governing valve control
AGCV	automatic governor valve control

Electrical Systems for Nuclear Power Plants, First Edition. Omar S. Mazzoni.
© 2019 by The Institute of Electrical and Electronic Engineers, Inc. Published 2019 by John Wiley & Sons, Inc.

AI	artificial intelligence
ALARA	as low as reasonably achievable
ALWR	advanced light-water reactor
APU	auxiliary power unit
APWR	advanced pressurized-water reactor
ARM	area radiation monitor
ARP	alarm response procedure
ASME	American Society of Mechanical Engineers
ASTM	American Society for Testing and Materials
ATWS	anticipated transient without scram
BAT	backup auxiliary transformer
BBS	battery backup system
BPV	boiler and pressure vessel
BV	bypass valve
BR	breeder reactor
BWR	boiling-water reactor
C/A	corrective action
CAA	Clean Air Act
CAC	carrier access code
CACS	containment atmosphere control system
CAD	computer-aided design
CAQ	condition adverse to quality
CAS	central alarm station
CB	containment building, control building
CCF	common-cause failure
CCFA	common-cause failure analysis/analyses
CCVT	capacitor voltage transformer
CCW	closed cooling water
CDBA	containment design-basis accident
C&I	control and instrumentation
Ci	Curie
CIAS	containment isolation actuation signal
CLB	current licensing basis
CMFA	common-mode failure analysis/analyses
COI	conflict of interest
COL	combined operating license, construction/operating license
CP	construction permit
	coolant pump
	critical power
cpm	count per minute
CPS	cathodic protection system
CPU	central processing unit

CR	condition report
	control room
CRD	control rod drive
CSCS	core standby cooling system
CSD	cold shutdown
CSF	critical safety factor
	critical safety function
CSP	Coalition for Safe Power
	conditional success probability
	containment spray pump
	core spray pump
CST	condensate storage tank
CT	cable test
	computerized tomography
	cooling tower
	combustion turbine
	current transformer
CTG	combustion turbine generator
CV	check valve
	containment vessel
	control valve
DA	deaerator
	dose assessment
DAS	data acquisition system
	disturbance analysis system
DBA	design-basis accident
DBD	design-basis document
DBE	design-basis events
DBLOCA	design-basis loss-of-coolant accident
DC	data control
	document control
	direct current
DCD	design control document
DCN	design change notice
	document change notice
DCPD	direct current potential drop
DCS	document control system
DCT	differential current transformer
DE	diagnostic evaluation
	dose equivalent
	double ended
DER	design electrical rating
	Deviation Event Report

DET	decomposition event tree
	diagnostic evaluation team
D&F	determination and findings
DFOS	diesel fuel oil system
DG	diesel-engine generator
	diesel generator
D/G	diesel generator
DGA	dummy guide assembly
DGAS	diesel generator auxiliary system
DGB	diesel generator building
DGCAIES	diesel generator combustion air intake and exhaust system
DGCWS	diesel generator cooling water system
DGSS	diesel generator starting system
DID	defense in depth
DMA	direct memory access
DMUX	distributed multiplex
DO	designated official
	digital output
DP	data processing
	differential pressure
D/P	differential pressure
dpm	decade per minute
	disintegrations per minute
DPMM	dewpoint moisture monitor
DPO	differing professional opinion
DPP	drip pan pot
DPR	developmental power reactor
DR	design review
D-T	deuterium–tritium
DTS	differential temperature switch
DVAL	degraded voltage analytical limits
DVM	digital voltmeter
DYNAL	dynamic analysis/analyses
DVR	dynamic voltage restoration
EA	enforcement action
	engineering assurance
	environmental assessment
EAP	emergency action plan
	Event Assessment Panel
EAPS	essential auxiliary power system
EC	eddy current
	emergency coordinator
	enforcement coordinator
	event category

ECC	emergency control center
	emergency core coolant
	emergency core cooling
ECCS	emergency core cooling system
ECCW	emergency core cooling water (system)
ECI	emergency coolant injection
	essential controls and instrumentation
ECS	emergency control station
ECT	eddy current test
	emergency cooling tower
ED&C	electrical distribution and control
EDG	emergency diesel generator
EDGS	emergency diesel generator sequencer
EDS	electrical distribution system
EDSFI	electrical distribution system functional inspection
E&I	electrical and instrumentation
ELMS	electrical load monitoring system
EMC	electromagnetic capability
EMI	electromagnetic interference
	engineering and manufacturing instruction
EMO	electric motor operated
EMOV	electromagnetically operated valve
EMP	electromagnetic pulse
EMT	electrical metallic tubing
EP	electric power
	emergency plan
	emergency planning
	emergency power
	emergency preparedness
	emergency procedure
E/P	electrical to pneumatic
EPA	electrical penetration assembly
EPM	electrical power monitoring
EPRI	Electric Power Research Institute
EPROM	electronic programmable read-only memory
EPS	electric power system
	emergency power system
EQ	environmental qualification
ERF	emergency response facility
ESF	engineered safety feature
FA	forced air
	fuel assembly
FDSA	facility description and safety analysis/analyses
FFD	fitness for duty

FOL	facility operating license
FP	fire protection
FPOL	full-power operating license
GBR	gas-cooled breeder reactor
GDC	general design criterion/criteria
GDP	gaseous diffusion plant
GIC	geomagnetically induced current
GIS	gas-insulated substation
GPS	global positioning system
GSU	generator stepup
GT-HTGR	gas-turbine high-temperature gas-cooled reactor
H/A	hand/automatic
HCE	human-caused error
HCV	hand control valve
HELBA	high-energy line break analysis/analyses
HER	human-error rate
HEU	highly enriched uranium
HFE	human factors engineering
HPC	health physics center
HPCI	high-pressure coolant injection
HPCS	high-pressure core spray
HPSI	high-pressure safety injection
HTGCR	high-temperature gas-cooled reactor
HV	high voltage
HVAC	heating, ventilation, and air conditioning
IA	industry application
	insertion approval
	instrument air
IAS	instrument air system
I&C	instrumentation and control
ICD	interface control diagram
	interface control document
	interface control drawing
ICEA	Insulated Cable Engineers Association
IDR	independent design review
IDV	independent design verification
IDVP	independent design verification program
IEEE	Institute of Electrical and Electronics Engineers
IAEA	International Atomic Energy Agency
ILRT	integrated leak rate test
IMS	in-core monitoring system
ISOs	Independent System Operators
INDE	inservice nondestructive examination

I/O	input/output
IPE	individual plant examination
	integrated plant evaluation
IPP	independent power producer
I/R	inductive to resistance
ISA	independent safety analysis
	Instrument Society of America
ISI	inservice inspection
IST	inservice test
LAN	local area network
LAR	license amendment request
LCO	limiting condition for operation
LCV	level control valve
	local control valve
LD	letdown
	lethal dose
LEP	local emergency protection
LER	licensee event report
LIMB	liquid metal breeder
LIRA	line resonance analysis
LLRW	low-level radioactive waste
LOCA	loss-of-coolant accident
LOOP	loss of off-site power
LOP	loss of power
LOV	loss of voltage
LPCI	low-pressure coolant injection
	low-pressure core injection
LPCIS	low-pressure coolant injection system
LV	low voltage
LRC	Locked Rotor Current
LVDT	linear variable differential transducer
LWR	light-water reactor
MCB	main control board
MCCB	molded-case circuit breaker
MCR	main control room
MEIS	minimum electrical isolation scheme
MEL	master equipment list
M/G	motor generator
MMI	man–machine interface
MPAI	maximum permissible annual intake
MPBB	maximum permissible body burden
MPC	maximum permissible concentration
MPFF	maintenance preventable functional failure

MPX	multiplexer
MSIV	main steam isolation valve
MTG	main turbine generator
MTTF	mean time to failure
MTTR	mean time to repair
MVAR	megavolt ampere reactive
MWe	megawatt electric
MWt	megawatt thermal
NCV	noncited violation
NE	normally energized
NEC	National Electrical Code
NEMA	National Electrical Manufacturers Association
NNI	nonnuclear instrumentation
NO	normally open
NOP	normal operating procedure
NOV	notice of violation
NPP	nuclear performance plan
	nuclear power plant
NRC	Nuclear Regulatory Commission
NSO	nuclear station operator
NSSS	nuclear steam supply system
	nuclear steam system supplier
OIC	online instrument and control
OL	operating license
	operating limit
	operator license
	overload (electrical)
OLD	one-line diagram
O&M	operation and maintenance
OS	operating system
PABX	private automatic branch exchange
PCIL	plant computer input list
PCIS	primary containment isolation system
PCIV	primary containment isolation valve
PDAS	plant data acquisition system
PFD	process flow diagram
PGCC	power generation control complex
P&I	piping and instrumentation
P&ID	piping and instrumentation diagram
PIS	pressure-indicating switch
	process instrumentation system
PLEX	plant life extension

PMS	performance measurement system
	plant monitoring system
	primary makeup system
	probable maximum surge
	program management staff
PORC	plant operations review committee
PORV	pilot-operated relief valve
	power-operated relief valve
ppm	part per million
PPS	plant protection system
	plant protective system
	plutonium product storage
	primary power system
pps	pulse per second
PRA	Paperwork Reduction Act of 1995
	plutonium recycle acid
	probabilistic risk analysis/analyses
	probabilistic risk assessment
PSAR	preliminary safety analysis report
PT	potential transformer
pu	per unit
PURPA	Public Utilities Regulatory Policy Act of 1978
PVC	polyvinyl chloride
QA	quality assurance
QC	quality control
RAB	reactor auxiliary building
RAM	radioactive material
	random access memory
	reliability, availability, and maintainability
RAW	risk assessment worth
RB	reactor building
RCA	Radiation Control Agency
	radiological controlled area
	reactor coolant activity
	root cause analysis
RCP	reactor cooling pump
RFI	radio frequency interference
	request for information
RFP	reactor feed pump
RG	regulatory guide
RHRS	residual heat removal system
rms	root mean square

RMS	radiation monitoring system
	radiological monitoring system
	regulatory manpower system
	remote manual switch
ROM	read only memory
rpm	revolutions per minute
RPS	radiation protection supervisor
	Reactor Program System
	reactor protection system
	regulatory performance summary
RSA	remote shutdown area
RTD	resistance temperature detector
	resistance temperature device
RTS	reactor trip system
RW	raw water
RWS	radwaste system
SA	safety analysis/analyses
SAE	site area emergency
SAR	safety analysis report
SAT	site access training
SBO	station blackout
SC	safety class
	secondary confinement
	site characterization
	site contingency
	speed controller
	surveillance compliance
S/D	shutdown
SER	Safety Evaluation Report
	Sequence of Events Recorder
SF6	sulfur hexafluoride
SSCs	structures, systems, and components
SGI	Safeguards Information
SI	safety injection
	security inspector
	special instruction
	surveillance inspection
	surveillance instruction
SIAS	safety injection actuation signal
SIS	safety injection signal
	safety injection system
	site-integrated schedule
	Special Inspection Services

SLD	shutdown logic diagram
SOE	sequence of events
SOP	standard operating procedure
SOV	solenoid-operated valve
SOW	statement of work
SP	sampling point
	security procedure
	set point
	special project
	special purpose
	suppression pool
	surveillance procedure
SPDS	safety parameter display system
SR	safety related
	safety rod
	source range
	summary report
	support reaction
	surveillance requirement
SRE	safety-related equipment
	sodium reactor experiment
SS	safe shutdown
SSC	structure, system, and component
SWS	service water system
TBS	turbine bypass system
T/C	thermocouple
TCCC	time current characteristic curve
TC/LD	thermocouple/lead detector
TD	theoretical density
	time delay
	transfer dolly
	turbine driven
TEDE	total effective dose equivalent
TEFC	totally enclosed fan cooled
TEG	thermoelectric generator
TEM	transmission electron microscope
TS	technical specification
TO	transmission system operator
TOC	time over current
TSSs	transmission system studies
UAT	unit auxiliary transformer
UPS	uninterrupted power supply
URI	unresolved item

UVTA	undervoltage trip attachment
VA	viability assessment
	vital area
	Volt Ampere
Vac	volts alternating current
VAR	volt amperes reactive
Vdc	volts direct current
VPC	volts per cell
VRLA	valve regulated lead acid
VSR	vertical slice review
WB	whole body
WD	waste disposal
XLPE	cross-linked polyethylene

Appendix 2

Definitions

(Collected from applicable IEEE Standards, IEC Electropedia " IEC 60050 - International Electrotechnical Vocabulary [www.electropedia.org], and US Nuclear Regulatory Commission publications)

Acceptable: Demonstrated to be adequate by the safety analyses of the station.
Actuated equipment: The assembly of prime movers and driven equipment used to accomplish a protective action.
Examples of prime movers are turbines, motors, and solenoids. Examples of driven equipment are pumps and valves.
Actuation device: A component or assembly of components that directly controls the motive power (e.g., electricity, compressed air, hydraulic fluid) for actuated equipment.
Examples of actuation devices are circuit breakers, relays, and pilot valves.
Administrative controls: Rules, orders, instructions, procedures, policies, practices, and designations of authority and responsibility.
Alternate ac source: An alternating current (ac) power source that is available to a nuclear power plant that meets special station blackout requirements.
Analysis: A process of mathematical or other logical reasoning that leads from stated premises to the conclusion concerning the qualification of an assembly or components.
Auxiliary supporting features: Systems or components that provide services (e.g., cooling, lubrication, energy supply) that are required for the safety systems to accomplish their safety functions.
Battery cell size: The rated capacity of a lead-acid cell or the number of positive plates in a cell.
Battery duty cycle: The loads a battery is expected to supply for specified time periods.
Battery equalizing charge: A prolonged charge, at a rate higher than the normal float voltage, to correct any inequalities of voltage and specific gravity that may have developed between the cells during service.

Electrical Systems for Nuclear Power Plants, First Edition. Omar S. Mazzoni.
© 2019 by The Institute of Electrical and Electronic Engineers, Inc. Published 2019 by John Wiley & Sons, Inc.

Battery full float operation: Operation of a DC system with the battery, battery charger, and load all connected in parallel and with the battery charger supplying the normal DC load plus any charging current required by the battery. (The battery will deliver current only when the load exceeds the charger output.)

Battery period: An interval of time in the battery duty cycle during which the load is assumed to be constant for purposes of cell-sizing calculations.

Battery-rated capacity (lead-acid): The capacity assigned to a cell by its manufacturer for a given discharge rate, at a specified electrolyte temperature and specific gravity, to a given end-of-discharge voltage.

Battery, valve-regulated lead-acid (VRLA): A lead-acid cell that is sealed with the exception of a valve that opens to the atmosphere when the internal gas pressure in the cell exceeds atmospheric pressure by a preselected amount. VRLA cells provide a means for recombination of internally generated oxygen and the suppression of hydrogen gas evolution to limit water consumption.

Battery vented: A battery in which the products of electrolysis and evaporation are allowed to escape freely to the atmosphere. These batteries are commonly referred to as "flooded."

Channel: An arrangement of components and modules required to generate a single protective action signal when required by a generating station condition. A channel loses its identity where single protective action signals are combined.

Class 1E: The safety classification of the electric equipment and systems that are essential to emergency reactor shutdown, containment isolation, reactor core cooling, and containment and reactor heat removal or that are otherwise essential in preventing significant release of radioactive material to the environment.

Containment: Engineered safety primary pressure boundary surrounding the reactor system pressure boundary, to provide a barrier to retain or prevent the release of radiation or hazardous contamination to the outside environment even under conditions of a reactor accident. May also be referred to as a confinement structure or vault.

Degraded voltage condition: A voltage deviation, above or below normal, to a level that, if sustained, could result in unacceptable performance of, or damage to, the connected loads and/or their control circuitry.

Design basis events: Postulated events used in the design to establish the acceptable performance requirements of the structures, systems, and components.

Design service conditions: The service conditions used as the basis for ratings and for the design qualification of electric equipment.

Design tests: Tests performed to verify that an electric equipment meets design requirements.

Detectable failures: Failures that can be identified through periodic testing or can be revealed by alarm or anomalous indication. Component failures that are detected at the channel, division, or system level are detectable failures.

Note: Identifiable but nondetectable failures are failures identified by analysis that cannot be detected through surveillance testing or cannot be revealed by alarm or anomalous indication.

Diesel generator unit: An independent source of standby electrical power that consists of a diesel-fueled internal combustion engine (or engines) coupled directly to an electrical generator (or generators); the associated mechanical and electrical auxiliary systems; and the control, protection, and surveillance systems.

Diesel generator unit continuous rating: The electric power output capability that the diesel generator unit can maintain in the service environment for 8760 h of operation per year with only scheduled outages for maintenance.

Diesel generator unit design load: That combination of electric loads (kW and kvar), having the most severe power demand characteristic, which is provided with electric energy from a diesel generator unit for the operation of engineered safety features and other systems required during and following shutdown of the reactor.

Diesel generator unit engine equilibrium temperature: The condition at which the jacket water and lube oil temperatures are both within ± 5.5°C (10°F) of their normal operating temperatures established by the engine manufacturer.

Diesel generator unit load profile: The magnitude and duration of loads (kW and kvar) applied in a prescribed time sequence, including the transient and steady-state characteristics of the individual loads.

Diesel generator unit service environment: The aggregate of conditions surrounding the diesel generator unit in its enclosure, while serving the design load during normal, accident, and postaccident operation.

Diesel generator unit short-time rating: The electric power output capability that the diesel generator unit can maintain in the service environment for 2 h in any 24 h period, without exceeding the manufacturer's design limits and without reducing the maintenance interval established for the continuous rating.

Diesel generator unit start signal: That input signal to the diesel generator unit start logic that initiates a diesel generator unit start sequence.

Division: The designation applied to a given system or set of components that enables the establishment and maintenance of physical, electrical, and functional independence from other redundant sets of components.

Documentation: Any written or pictorial information describing, defining, specifying, reporting, or certifying activities, requirements, procedures, or results.

Electrical penetration assembly: An assembly of insulated electric conductors, conductor seals, module seals (if any), and aperture seals that provides the passage of the electric conductors through a single aperture in the nuclear containment structure, while providing a pressure barrier between the inside and the outside of the containment structure. The assembly may include optical fibers and fiber seals. The electric penetration assembly includes terminal (junction) boxes, terminal blocks, connectors and cable supports, and splices that are designed and furnished as an integral part of the assembly.

Electrical penetration assembly double aperture seal: Two single aperture seals in series.

Electrical penetration assembly double optical fiber seal: Two single optical fiber seals in series.

Electrical penetration assembly double single aperture seal: A single seal between the containment aperture and the electric penetration assembly.

Electrical penetration assembly single optical fiber seal: One single optical fiber seal to create a single pressure barrier seal between the inside and the outside of the containment structure along the axis of the optical fiber.

Electrical penetration assembly single electric conductor seal: A mechanical assembly arranged in such a way that there is a single pressure barrier seal between the inside and the outside of the containment structure along the axis of the electric conductor.

Engineered safety features: Features of a unit, other than reactor trip or features used only for normal operation that are provided to prevent, limit, or mitigate the release of radioactive material.

Environmental conditions: Physical service conditions external to the electric equipment such as ambient temperature, pressure, radiation, humidity, vibration, chemical or demineralized water spray and submergence expected as a result of normal operating requirements, and postulated conditions appropriate for the design basis events applicable to the electric equipment.

Execute features: The electrical and mechanical equipment and interconnections that perform a function, associated directly or indirectly with a safety function, upon receipt of a signal from the sense and command features. The scope of the execute features extends from the sense and command features output to and including the actuated equipment-to-process coupling.

Generic design: A family of equipment units having similar materials, manufacturing processes, limiting stresses, design, and operating principles, that can be represented for qualification purposes by a representative unit(s).

Independence: The state in which no mechanism exists by which any single design basis event can cause redundant equipment to be inoperable.

Installed life: The interval of time from installation to permanent removal from service, during which the electric equipment or assembly is expected to perform its required function(s).

Note: Components of the assembly may require periodic replacement; when the installed life of such components is less than the installed life of the assembly

Isolating device: A device in a circuit that prevents malfunction in one section of a circuit from causing unacceptable influences in other sections of the circuit or in other circuits.

Load group: An arrangement of buses, transformers, switching equipment, and loads fed from a common power supply within a division.

Load profile: The magnitude and duration of loads (kW and kvar) applied in a prescribed time sequence, including the transient and steady-state characteristics of the individual loads.

Locked rotor torque: The torque at the instant of motor start.

Loss of off-site power: The loss of all power from connections to the grid.

Loss of voltage condition: A voltage reduction to a level that results in the immediate loss of equipment capability to perform an intended function.

Margin: The difference between the most severe design service conditions and the conditions used in the design qualification to account for normal variations in commercial production of equipment and reasonable errors in defining satisfactory performance.

Module: Any assembly of interconnected components that constitutes an identifiable device, instrument, or piece of equipment. A module can be disconnected, removed as a unit, and replaced with a spare unit. It has definable performance characteristics that permit it to be tested as a unit. A module could be a card, a drawout circuit breaker, or other subassembly of a larger device, provided it meets the requirements of this definition.

Nuclear power generating station (station): A plant where electric energy is produced from nuclear energy by means of suitable apparatus. The station may consist of one or more generating units.

On-site power supply: The emergency ac and DC power supplies, including their associated electrical distribution systems, classified as Class 1E.

Power sources: The electrical and mechanical equipment and interconnections necessary to generate or convert power.

Note: Electric power source and power supply are interchangeable within the context of this document.

Preferred power supply (also called "off-site" power supply): The power supply from the transmission system to the Class 1E distribution system that is preferred to furnish electric power under accident and postaccident conditions.

Programmable digital computer: A device that can store instructions and is capable of executing a systematic sequence of operations performed on data that is controlled by internally stored instructions.

Protection system: The part of the sense and command features involved in generating the signals used primarily for the reactor trip system and engineered safety features.

Protective action: The initiation of a signal within the sense and command features, or the operation of equipment within the execute features, to accomplish a safety function.

Pump runout: A pump flow condition in which the pump is operating beyond its design point due to a reduction in the system head. As a result, the pump motor's brake-horsepower and full load current demand may be increased.

Qualified diesel generator unit: A diesel generator unit that meets the qualification requirements of the regulatory standards.

Qualified life: The period of time, prior to the start of a design basis event, for which the equipment was demonstrated to meet the design requirements for the specified service conditions.

Note: At the end of the qualified life, the equipment shall be capable of performing the safety function(s) required for the postulated design basis and postdesign basis events.

Qualified life test: Tests performed on preconditioned test specimens to verify that an electric equipment will meet design requirements at the end of its qualified life.

Redundant equipment or system: A piece of equipment or a system that duplicates the essential function of another piece of equipment or system to the extent that either may perform the required function regardless of the state of operation or failure of the other.

Note: Redundancy can be accomplished by using identical equipment, equipment diversity, or functional diversity.

Safety class structures: Structures designed to protect Class 1E equipment against the effects of design basis events.

Safety function: One of the processes or conditions (e.g., emergency negative reactivity insertion, postaccident heat removal, emergency core cooling, postaccident radioactivity removal, containment isolation) essential to maintain plant parameters within acceptable limits established for a design basis event.

Note: A safety function is achieved by the completion of all required protective actions by the reactor trip system and the engineered safety features, or both, concurrent with the completion of all required protective actions by the auxiliary supporting features.

Safety group: A given minimal set of interconnected components, modules, and equipment that can accomplish a safety function.

Note: A safety group may include one or more divisions. In a design where each division can accomplish a safety function, each division is a safety group. However, a design consisting of three 50% capacity systems separated into three divisions would have three safety groups; any two out of three divisions are required to be operating to accomplish the safety function.

Safety system: A system that is relied upon to remain functional during and following design basis events to ensure (1) the integrity of the reactor coolant pressure boundary, (2) the capability to shut down the reactor and maintain it in a safe shutdown condition, or (3) the capability to prevent or mitigate the consequences of accidents that could result in potential offsite exposures comparable to the 10CFR Part 100 guidelines.

Sense and command features: The electrical and mechanical components and interconnections involved in generating the signals associated directly or indirectly with the safety functions. The scope of the sense and command features extends from the measured process variables to the execute features input terminals.

Significant: Demonstrated to be important by the safety analysis of the station.

Service conditions: Environmental, power, and signal conditions expected as a result of normal operating requirements, expected extremes in operating requirements, and postulated conditions appropriate for the applicable design basis events.

Service factor (general-purpose alternating current motor): A multiplier that, when applied to the rated power, indicates a permissible power loading that may be carried under the conditions specified for the service factor.

Severe accident conditions: Postulated events with environmental conditions more severe than used in the design to establish the acceptable performance requirements of the electrical equipment.

Standby power supply: The power supply that is selected to furnish electric energy when the preferred power supply is not available.

Station blackout: The complete loss of alternating current (ac) electric power to the essential and nonessential switchgear buses in a nuclear power plant (i.e., loss of off-site electric power system concurrent with turbine trip and unavailability of the on-site emergency ac power system). Station blackout does not include the loss of available ac power to buses fed by station batteries through inverters or by alternate ac sources.

Surveillance: The determination of the state or condition of a system or subsystem.

Unit: A nuclear steam supply system and its associated turbine generator, auxiliaries, and engineered safety features.

Verification and validation: The process of determining whether the requirements for a system or component are complete and correct, the products of each development phase fulfill the requirements or conditions imposed by the previous phase, and the final system or component complies with specified requirements.

Transmission system studies: These studies are performed to demonstrate that the preferred power supply is not degraded below a level consistent with the availability goals of the plant as a result of contingencies such as any non-simultaneous events.

Valve actuator duty cycle: The duty cycle of a valve actuator that consists of the number of valve strokes needed for its intended service. Plant operation or testing may require successive stroking of the valve.

Valve actuator duty cycle time: The time required for a valve to complete one stroke multiplied by the duty cycle.

Valve actuator motor nominal current: The current that the motor is rated to carry for the duration of its rated duty, commonly known as the motor nameplate current.

Valve actuator motor nominal torque: The torque that corresponds to the valve actuator nominal current.

Valve stroke: The travel of a valve from a fully open to a fully closed position or vice versa.

Valve (motor-operated) anticipated overloads: Those values of motor current in excess of motor nameplate rating that are expected to be realized during valve actuation due to the varying torque nature of the load.

Index

Electrical Systems for Nuclear Power Plants, First Edition. Omar S. Mazzoni.
© 2019 by The Institute of Electrical and Electronic Engineers, Inc. Published 2019 by John Wiley & Sons, Inc.